Pittsburgh in 50 Maps

Pittsburgh in 50 Maps

by Stentor Danielson

Belt Publishing
Pittsburgh, PA

Copyright © 2025 by Belt Publishing

All rights reserved. This book or any portion thereof may not be reproduced or used in any manner whatsoever without the express written permission of the publisher except for the use of brief quotations in a book review.

First edition, 2025
ISBN: 978-1-953368-85-0

Belt Publishing
6101 Penn Avenue
Suite 201
Pittsburgh PA 15206
www.beltpublishing.com

Cover design by David Wilson

I will see the city poured rolling down the mountain valleys like slag, and see the city lights sprinkled and curved around the hills' curves, rows of bonfires winding.

–Annie Dillard, *An American Childhood*, 1987

Calvin: "I wonder where we go when we die."
Hobbes: "Pittsburgh?"
Calvin: "You mean if we're good or if we're bad?"

–Bill Watterson, *Calvin and Hobbes*, 1985

Contents

Introduction 9

Section One: Situating the City

The Center of It All 12
City of Hills 15
City of Four Rivers? 16
Jaödeogë': A City on Indigenous Land . . . 19
A Growing City 20
City Boundaries. 23
Shrinking and Sprawling. 24
Coal City. 27
Sister Cities 28
Postal Pittsburgh 31
Famous Firsts 32
Pittsburgh With an *h* 35
Yinz N'At. 36

Section Two: Getting around the City

Highways, Tunnels, and Belts. 41
Allegheny, Monongahela,
and Ohio Watersheds 42
Islands of Pittsburgh. 45
City of Bridges 46
The Pittsburgh Left 49
Underground Railroad 50
Pittsburgh Steps 53
Ride the Trolley 54
Bike Lanes. 57
Pittsburgh Marathon. 58
The Parking Spot Outside the
Evergreen Cafe on Penn Avenue 61

Section Three: Communities and Neighborhoods

A City of Neighborhoods	65
Mount Oliver	66
Gentrification	69
Old Houses and Lead Paint	70
Children and Seniors	73
Race and Ethnicity	74
Tree Cover	77
Supermarkets and Food Deserts	78
Chinatown	81
Immigrant Population	82
LGBTQ+Burgh	85
Jewish Pittsburgh	86
Catholic Pittsburgh	89
Muslim Pittsburgh	90
Mexican War Streets	93
Blue Dot in a Red Sea	94
To Outer Space	97

Section Four: Places and People in the City

The Cathedral of Learning	101
Mister Rogers' Neighborhood	102
City of Dinosaurs	105
Carnegie's Legacy	106
Steel City	109
Eds and Meds	110
Pets in the Burgh	113
Air Quality	114
Fracking and Sunning	117
Rachel Carson	118
August Wilson	121
The Sporting Life	122
The Immaculate Reception	125

Sources	127
Acknowledgments	135
About the Author	136

Editor's Note
Similar to the other titles in this series, this book should really be called *Pittsburgh in (About) 50 Maps*. Depending how you count—and what you think qualifies as "one" map—you'll find between fifty and one hundred maps collected here. They all offer unique insights into the history, demographics, and culture of a great American city. But fifty's been a nice round number for us with this series, so we're sticking with it.

Introduction

Steel City. City of Bridges. Hell with the Lid Off. City of Champions. Whatever you call it, Pittsburgh's distinctive character is undeniable.

The confluence of the Allegheny, Monongahela, and Ohio Rivers was always destined to be a key crossroads of the North American continent. Seneca and Shawnee people canoed up and down these waterways for thousands of years before the French and English showed up to build Fort Duquesne and Fort Pitt in the middle of the eighteenth century. Not long after that, more white settlers came to the area to build and work in the coal, iron, glass, and steel industries, eventually learning to organize for better pay and more humane working conditions. Black Americans migrated up from the South, demanding freedom and equal treatment under the law. Today, people still come from all over the world to make their home here, as you'll know if you've ever visited the "Himalayan Highway" (Route 51) for some momos made by Bhutanese refugees. When the steel mills closed in the 1980s, the city hit some economic doldrums: people left, lots became vacant, and infrastructure crumbled. But Yinzers have refused to give up on the city. Tech companies like Google have brought new jobs to the region, and housing remains relatively affordable. Local environmental and racial justice advocates continue to point out the ways that Pittsburgh falls short of its reputation as the country's "most livable city," but you wouldn't fight so hard to improve a city if you didn't believe in its potential.

In the maps collected here, you'll find Pittsburgh's lovable quirks as well as its serious social and environmental issues. We've got Mister Rogers' Neighborhood, and we've also got air pollution. We've got that annoying car parked outside the Evergreen Cafe on Penn Avenue, and we've also got the undeniable racial segregation that still marks the city today. We've got the Pittsburgh Left, and we've got people losing their homes to gentrification.

Though I was born in Ohio, my parents moved us to Forest County, Pennsylvania, when I was only a year old, so I've been in Pittsburgh's orbit from a very young age. When I was nine, we moved again, to eastern Pennsylvania, and from there, I went to college in Upstate New York and to graduate school in central Massachusetts before ending up in Arizona with my partner. It was in the arid Southwest that I started to grasp just how special Pittsburgh was. I was constantly meeting Yinzers who had moved to Arizona—so many of them that in 2009, when the Steelers played the hometown Cardinals in Super Bowl XLIII, the state was split on who to support.

That same year, I landed a job teaching geography at Slippery Rock University, and my partner and I decided to move back east. The college is an hour north of Pittsburgh, but we decided to live in the city to take advantage of everything it has to offer (a special shout-out to the improv comedy scene, which my partner dove into right away). We lived in the North Side for a decade before housing costs led us just across the city line into Bellevue and then a bit farther west, into Avalon.

Moving back and living here as an adult gave me some new insights into the city. I hope this book captures some of the most interesting things I've discovered about Pittsburgh and its surrounding region over the past fifteen years. "Situating the City" examines Pittsburgh's unique location, showing how it has been shaped by local factors such as its geology and Indigenous history, how it has grown to its present boundaries, and why this Pittsburgh is the only one in the country that ends with an *h*. "Getting Around the City," showing how people move around and through the city, highlights unique features like the city's nearly three hundred bridges and eight hundred sets of public stairs, as well as that infamous bit of Steel City courtesy known as the Pittsburgh Left. "Communities and Neighborhoods" looks at how the city's ninety-five neighborhoods get their distinctive character from factors like their ethnic makeup, religious institutions, tree cover, and housing stock and does a deep dive into a few neighborhoods of note, such as the section of the North Side where the streets are named for battles in the Mexican-American War. Finally, "Places and People in the City" shows how Pittsburgh is being shaped by contemporary issues like fracking and the Highmark-UPMC rivalry, and traces the legacy of famous residents like Fred Rogers, Rachel Carson, and August Wilson.

Every map tells a story, and the maps in this book tell some of the stories that I think are particularly interesting about Pittsburgh. As you peruse them, I hope they'll also make you consider all the Pittsburgh stories they don't tell. And I hope this book inspires you to go find—or make—maps that tell the stories about this wonderful, complex city that matter most to you.

SECTION ONE
Situating the City

The Center of It All
City of Hills
City of Four Rivers?
Jaödeogë': A City on Indigenous Land
A Growing City
City Boundaries
Shrinking and Sprawling
Coal City
Sister Cities
Postal Pittsburgh
Famous Firsts
Pittsburgh With an *h*
Yinz N'At

The Center of It All

There are lots of ways to describe Pittsburgh's location—Southwestern Pennsylvania; the confluence of the Allegheny, Monongahela, and Ohio Rivers; a three-hour drive from Cleveland or Buffalo. Geographers like to locate things by latitude and longitude coordinates. By that measure, the Point, where the three rivers come together at the symbolic heart of the city, is at N 40° 26' 31" W 80° 00' 46". Knowing this, we can create a map of the whole world in which every place is the correct distance and direction from Pittsburgh. (Geographers call this an azimuthal equidistant projection).

Within one thousand miles of Pittsburgh, you can reach Miami, Boston, or Kansas City. Going two thousand miles can bring you to Phoenix, Puerto Rico, or Iqaluit in northern Canada. A journey of five thousand miles as the crow flies would reach Honolulu, Hawaii; Santiago, Chile; Kyiv, Ukraine; or Monrovia, Liberia.

The outer edge, on this map, represents Pittsburgh's antipode—the point exactly on the opposite side of the earth, some 12,450 miles away. Pittsburgh's antipode is somewhere in the Indian Ocean, about one thousand miles southwest of Perth, Australia.

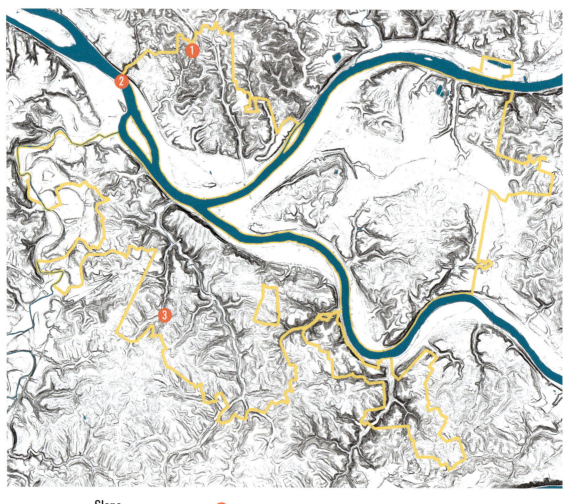

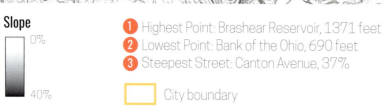

Slope 0% – 40%

1. Highest Point: Brashear Reservoir, 1371 feet
2. Lowest Point: Bank of the Ohio, 690 feet
3. Steepest Street: Canton Avenue, 37%

☐ City boundary

City of Hills

Geologically, Pittsburgh is situated on the Allegheny Plateau, a broad landform making up most of Western Pennsylvania. The rocks under our feet—sandstone, shale, coal, and limestone—were laid down some three hundred million years ago, during the appropriately named Pennsylvanian Period. At that time, this region was a shallow sea that received sediment washed down from the newly forming Appalachian Mountains to the east. Over time, the plateau was pushed upward as North America collided with Africa and Europe to form the ancient supercontinent of Pangea.

In the millions of years since, rivers and streams have been eroding valleys down into the plateau, sometimes filling the bottoms with new sediment. The result is a landscape made up of flat hilltops and valley bottoms separated by steep hillsides. The city's highest point is the Brashear Reservoir atop Observatory Hill, 1,371 feet above sea level. The lowest point is along the north bank of the Ohio River, where it meets Jacks Run on the city's border. Canton Avenue in Beechview is the steepest street not only in the city but in the entire country, with a slope of 37 degrees.

The prevalence of steep slopes makes Pittsburgh especially vulnerable to landslides, especially after heavy rains make the soil soft and "slippy." A major landslide in Duquesne Heights forced the temporary closure of Route 51 in 2018. Because climate change is expected to increase rainfall in the area, this risk will only become greater in the future. With the help of a $10 million grant from FEMA, the city recently began a major project to stabilize hillsides and buy out threatened homes in Mount Washington and other steep areas in the southern part of the city.

City of Four Rivers?

Pittsburgh is famously known as the City of Three Rivers. But rumors abound of a mysterious fourth river, which runs below the Point and feeds the famous fountain. In fact, that fourth underground waterway is real, though it's not the kind of river you would usually imagine. You can't ride a boat down it or fish off its banks. It's the aquifer underlying the city.

During the last ice age, which took place from about 2.5 million to 12,000 years ago, sediment washed downstream from the glaciers covering the northern part of Pennsylvania and was deposited along the river valleys of Pittsburgh, where it formed a layer of porous sand and gravel fifty feet thick. The water below, which stays at a steady fifty degrees, can be tapped by wells to provide drinking water or to help cool buildings. (Alas, contrary to popular belief, the fountain at the Point is fed by the city's regular water system, and not directly by this "fourth river.")

Another type of hidden waterway in the city is the numerous streams that have been buried by development. Historic maps show a variety of waterways that are now lost. For example, a 1795 map shows two large ponds downtown: one along Liberty Avenue between Fourth and Fifth Avenue, and another stretching from Fourth and Smithfield to Strawberry Way and Grant. Sakes Run used to flow along the southern edge of the Hill District and into the Monongahela near the Liberty Bridge. Four Mile Run connected Panther Hollow to the Monongahela. An unnamed stream traversed the North Side as well, ending near the Fort Duquesne Bridge. All of these have been covered or filled in over decades of development, but during heavy rains, flooding can still occur along the old stream courses.

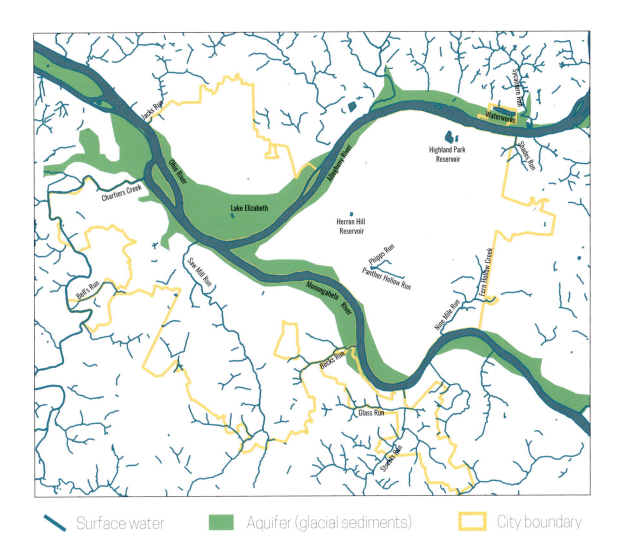

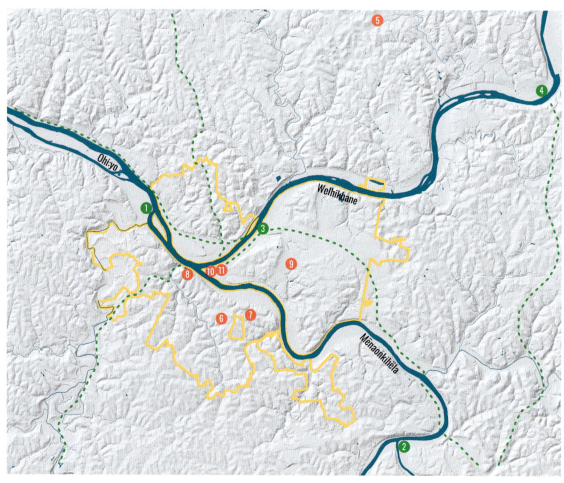

Historic Sites
1. Indian Mound (Adena)
2. Queen Aliquippa's settlement (Seneca)
3. Shannopin's Town (Lenape)
4. The-we-gi-la (Shawnee)

- - - Path

☐ City boundary

Contemporary Locations
5. Council of Three Rivers American Indian Center
6. COTRAIC Head Start
7. COTRAIC Early Head Start
8. Statue of Seneca leader Guyasuta meeting George Washington
9. Alcoa Foundation Hall of American Indians (Carnegie Museum of Natural History)
10. Fort Pitt Museum
11. Grave of Shawnee leader Red Pole

Jaödeogë': A City on Indigenous Land

Prior to European settlement, the Pittsburgh area was inhabited by the Onödowá'ga: (Seneca), Šaawanwaki (Shawnee), Eriehronon (Erie), Lenni Lenape (Delaware), and Gandastogue (Susquehannock) nations. The Seneca called the site Jaödeogë', meaning "between two rivers," while the Shawnee called it Hothawikamiki, meaning "by the yellow water." When Europeans came to the area in the eighteenth century, they encountered a Lenape village at Shannopin's Town (modern-day Lawrenceville), a Seneca village led by Queen Aliquippa (modern-day McKeesport), and a Shawnee settlement at The-we-gi-la (modern-day Springdale). A burial mound from the Adena culture, dated between 1000 and 1500 years ago, sat on the bank of the Ohio River in McKees Rocks until it was destroyed by inept archaeological excavations in the late 1800s.

Pittsburgh's rivers bear Indigenous names. The Seneca and Lenape people view the Ohio and Allegheny as a single river and call it the "good river" in their respective languages. The English names come from those Indigenous names, Seneca Ohi:yo and Lenape Welhikhane. Monongahela comes from the Lenape name Mënaonkihëla, meaning "where the banks cave in." A tributary of the Allegheny near the Waterworks Mall was (like many places around the country) named with a slur for Native women until 2022, when it was renamed Sycamore Run.

According to the 2020 Census, Allegheny County was home to some 1,049 people who identified themselves as American Indian or Alaska Native alone, and another 11,081 who were American Indian or Alaska Native along with one or more other races. This includes members of dozens of different nations, including Blackfeet, Cherokee, and various Indigenous Mexican peoples. Indigenous Pittsburghers face particular challenges due to their lack of formal recognition as a community and separation from traditional homelands. The Council of Three Rivers American Indian Center in Dorseyville was founded in 1970 to support this community. The center hosts various cultural activities, including an annual pow-wow, and provides economic support and social and educational programs for children and elders.

A Growing City

When Pittsburgh was first incorporated in 1794, it consisted of just the flat triangle of land that makes up Downtown today. Over the next two centuries, this little 0.6 square-mile area would grow into the 58.4 square-mile city we know now.

The city's first annexations, in the 1830s and 1840s, incorporated the Strip District, Bluff, and Lower Hill areas. Big change came in 1868, when most of the area comprising the city's eastern portion, between the Allegheny and Monongahela rivers, was added in one fell swoop. Four years later, the city crossed a river for the first time to absorb the city of Birmingham, which now makes up Mount Washington and the South Side.

Another major change came in 1907, when Pittsburgh annexed the much wealthier Allegheny City to the north. This annexation was the subject of a referendum, which was voted down within Allegheny but prevailed in the combined area. The US Supreme Court ruled in Pittsburgh's favor, and Allegheny became the North Side.

Further areas were added to the city in the south and west during the first few decades of the twentieth century. The last major area to be added was Hays Woods, annexed from Baldwin Borough in 1951. In recent years, another proposed expansion would have added the borough of Wilkinsburg to the city. Some officials in the borough thought annexation would help resolve problems with a declining tax base and urban blight, but a Commonwealth Court decision in 2023 put the kibosh on the plan.

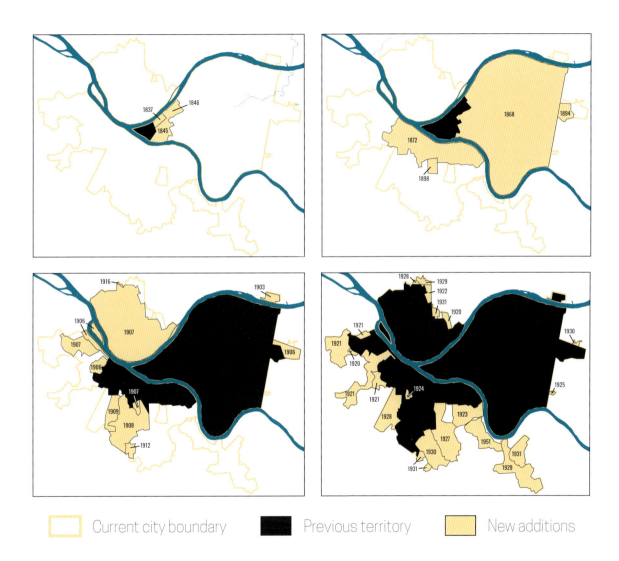

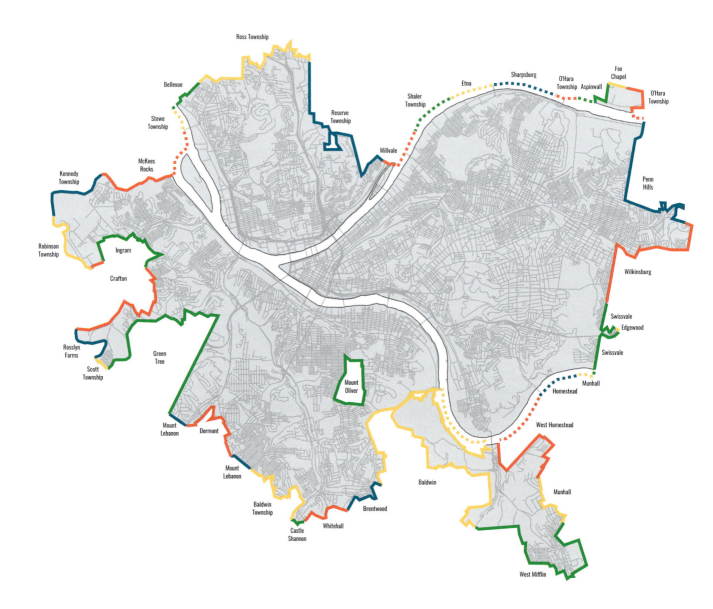

City Boundaries

Pittsburgh is a city defined more by its center than its edges. The city sits at the heart of Allegheny County, Pennsylvania's largest county by land area. And at the heart of the city is the confluence of the Allegheny and Monongahela, where they form the Ohio. At its edges, Pittsburgh abuts thirty-five other municipalities, but these borders rarely follow any major natural features. In many spots, the border slices through what would appear to be a continuous neighborhood.

Starting on the north bank of the Ohio, at the mouth of Jacks Run, the city is bordered by the borough of Bellevue, then Ross Township, then Reserve Township. From there, the border follows the Allegheny, with the city to the south and Millvale, Shaler Township, Etna, and Sharpsburg on the north bank. A small bit of Pittsburgh on the north bank around the Waterworks Mall is bordered by Aspinwall, Fox Chapel, and O'Hara Township. As the city border cuts across from the Allegheny to the Monongahela, it separates Pittsburgh from Penn Hills, Wilkinsburg, Edgewood, and Swissvale. The Monongahela forms the city boundary for a while, with Munhall, Homestead, and West Homestead on the south bank. Then the city crosses over, touching West Homestead and Munhall by land as well as West Mifflin and Baldwin. A sliver of Baldwin curls around the Hays neighborhood to separate it from the river. Brentwood, Whitehall, Castle Shannon, Baldwin Township, Mount Lebanon, Dormont, and Green Tree border Pittsburgh to the south. Coming around to the west side, the city shares a border with Scott Township, Carnegie, Rosslyn Farms (along Chartiers Creek), Crafton, and Ingram. Chartiers Creek forms the border with Robinson Township, Kennedy Township, and McKees Rocks, where it flows into the Allegheny. Stowe Township occupies the south bank of the Ohio as the border returns to Bellevue.

Careful readers will notice that's only thirty-four municipalities. The thirty-fifth is the borough of Mount Oliver, which is an enclave completely surrounded by Pittsburgh. Mount Oliver gets its own map later in this book.

Shrinking and Sprawling

Over the last half century, the population of Pittsburgh and its surrounding area has been declining. This decline has not been even across all areas, though; the city and the old industrial towns along the rivers have lost population more dramatically, while suburban and rural surrounding communities have experienced slower declines or even growth. The result has been a more sprawling metropolis, with the people of the region spread out farther from the Point.

The city reached its peak population at 676,806 in 1950. By 2020, only 302,971 people lived inside the city limits—a decline of 55 percent. Outside the city, Allegheny County's population peaked at 1.63 million in 1960; in 2020, it stood at 1.25 million, a decline of only 33 percent. Even further out, the population of the eight-county metropolitan area declined from 2.77 million to 2.37 million between 1960 and 2020—a drop of just 14 percent.

A 2014 report from Smart Growth America ranked Pittsburgh as the 132nd least sprawling city out of 221 in the country, more sprawling than Buffalo, Detroit, and even Los Angeles, but less than Cleveland. The sprawl index used in that report considered population density, land use mix, concentration of jobs in the city center, and street connectivity.

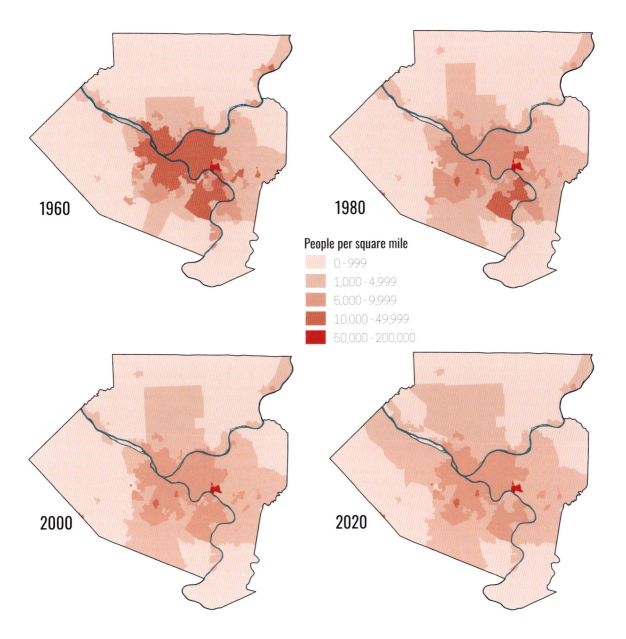

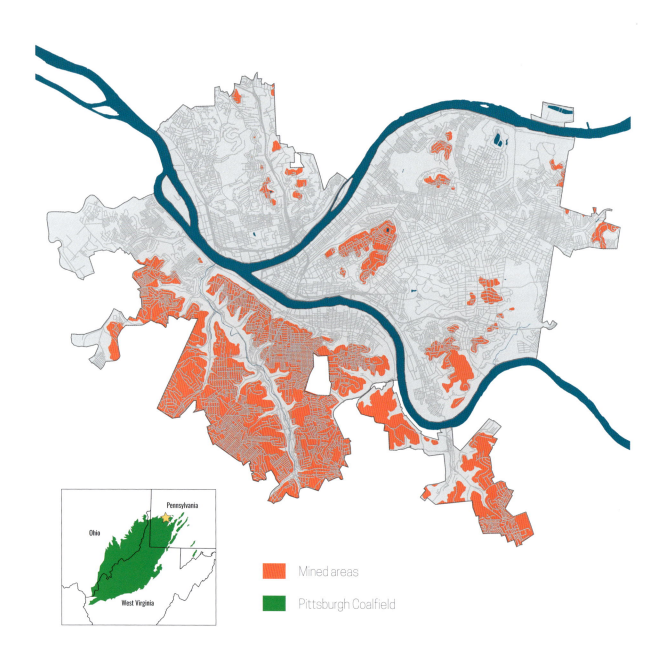

Coal City

Pittsburgh is known as the Steel City, but to make steel you need coal. Conveniently, the Pittsburgh Coalfield stretches across southwestern Pennsylvania, providing a rich source of bituminous coal. This coal seam, formed from ancient forests that died and were buried in coastal swamplands some three hundred million years ago, is about six feet thick. River valley erosion eventually exposed the coal layer on the side of Mount Washington, which early European settlers originally dubbed "Coal Hill." Nearly all coal in the region was extracted from underground "room and pillar" mines rather than open surface mines. Some sixty separate mines operated at one time within the city.

Coal mining in the area declined in the early twentieth century due to changes in steel-making technology and shifts to other energy sources for domestic and industrial power. But the history of mining has left many areas of the city—forty-two thousand homes, according to Department of Environmental Protection estimates—at risk of collapse or damage to building foundations. If you buy a house in one of the undermined areas, it's good to get insurance just in case. The decline in coal use also contributed to big improvements in the city's air quality, but water quality can still suffer from acid mine drainage as groundwater seeps through abandoned mines.

Sister Cities

Pittsburgh has sister city agreements with twenty-one other cities around the world. Most of these partners are in Europe, but several can be found in Asia and Latin America as well. Many—like Bilbao, Spain; Pohang, Korea; and Sheffield, England—are also known for their steel industries. These sister city relationships have helped to foster cultural exchange and economic cooperation, such as a 2002 exhibit called "The Bilbao Effect" at the Carnegie Science Center.

Some sister city agreements go back decades, but starting in 2017, then-mayor Bill Peduto made a big push to reinvigorate the program. One of the city's newest partnerships is with Glasgow, Scotland, focusing on environmental sustainability. The creation of this relationship coincided with a trip by Peduto to the COP26 United Nations climate change conference in Glasgow in 2021.

One of Pittsburgh's sister cities gained international notoriety starting in 2019, when the covid-19 pandemic began in Wuhan, China. The two cities had first connected in 1982 over their shared situations as industrial hubs on major rivers looking to reinvent themselves for the post-industrial era. Several hundred people from Wuhan's Hubei Province live in Pittsburgh, and a longstanding agreement had teachers from Wuhan University providing Chinese lessons at the University of Pittsburgh. These connections encouraged Pittsburgh's Chinese community to move quickly to send aid to Wuhan as it suffered the early ravages of the coronavirus and the ensuing lockdowns.

Another sister city grabbing headlines is Donetsk, Ukraine, one of the eastern cities targeted in Russia's 2022 invasion. In the 1980s and 90s the relationship with Pittsburgh led to a youth hockey tournament and the creation of a Ukrainian documentary called "Pittsburgh the City of Hope." Though Pittsburgh has a large Ukrainian population, this particular sister city relationship has been quiet in recent years.

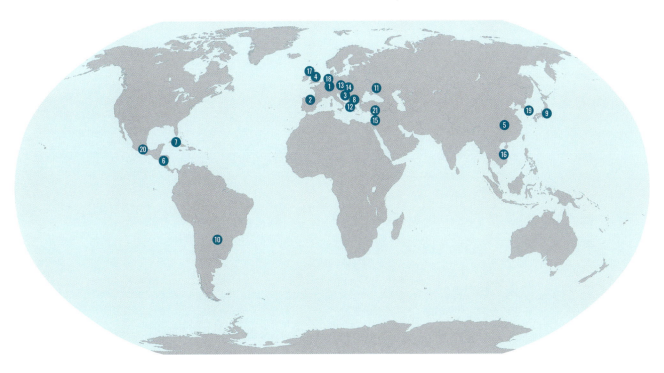

1. Saarbrucken, Germany (1956)
2. Bilbao, Spain (1970)
3. Zagreb, Croatia (1979)
4. Sheffield, England (1980)
5. Wuhan, China (1982)
6. San Isidro, Nicaragua (1987)
7. Matanzas, Cuba (1993)
8. Sofia, Bulgaria (1993)
9. Saitama City, Japan (1997)
10. Fernando de la Mora, Paraguay (1999)
11. Donetsk, Ukraine (1999)
12. Skopje, North Macedonia (2000)
13. Ostrava, Czech Republic (2001)
14. Presov, Slovakia (2002)
15. Karmiel & Misgav, Israel (2006)
16. Da Nang, Vietnam (2008)
17. Glasgow, Scotland (2020)
18. Dortmund, Germany (2022)
19. Pohang, Korea (2023)
20. Naucalpan, Mexico (date unknown)
21. Gaziantep, Turkey (date unknown)

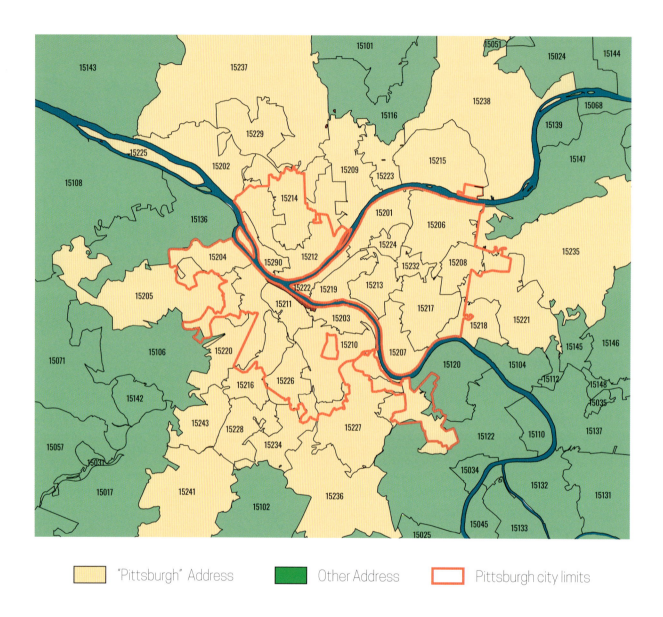

Postal Pittsburgh

There are a variety of ways to define a city's boundaries. I often say I live in Pittsburgh even though my house is technically in the borough of Avalon. While some folks may quibble with my claim to residency since I don't live within the official city limits, the US Postal Service backs me up. A number of ZIP codes outside of the city officially have Pittsburgh addresses. People living as far south as Pleasant Hills or as far north as Ingomar can put "Pittsburgh" on their mail. Conversely, residents of some parts of the New Homestead neighborhood of Pittsburgh, if their ZIP code is 15210, have "Homestead" addresses.

USPS establishes ZIP codes based on mail delivery routes, grouping together areas that can be easily served by the same mail carriers. The official city designation is based on the post office from which the mail is dispatched. My home in 15202 lies outside the city limits, but our mail comes via the distribution center on California Avenue in the North Side. Thus: Pittsburgh.

Famous Firsts

Pittsburgh has a long reputation for innovation, and over the years the city has been the first to introduce a variety of new technologies, ideas, and institutions to the world. Many of these "firsts" related to the city's industrial growth, such as the 1845 construction of the world's first wire cable suspension aqueduct bridge, which carried the Pennsylvania Main Line Canal across the Allegheny River into downtown, opening the city to barge traffic from Philadelphia. And if you're blowing up your group chat with emojis, you may be interested to know that their predecessor, the sideways typographic emoticon, was first created right here at Carnegie Mellon University. The Pittsburgh Pirates have been especially prolific in claiming firsts, playing the first World Series in 1903, making their home in the first baseball stadium in 1909, and playing the first World Series night game in 1971.

1. First newspaper west of the Alleghenies, the *Pittsburgh Gazette*, 1789
2. First wire cable suspension aqueduct bridge in the world, Pennsylvania Main Line Canal crossing the Allegheny River, 1845
3. First national convention of Federation of Organized Trades and Labor Unions, the precursor to the American Federation of Labor (AFL), 1881
4. First Major League Baseball World Series, games 4–7 between the Boston Americans and the Pittsburgh Pirates, Exposition Field, 1903
5. First dedicated motion picture theater in the world, the Nickelodeon, 1905
6. First baseball stadium in the US, Forbes Field, 1909
7. First gas station in the US, Gulf Refining Company, 1913
8. First polio vaccine, developed by Jonas Salk at the University of Pittsburgh, 1954
9. First public television station in the US, WQED, 1954
10. First retractable dome sports arena in the world, Mellon Arena, 1961
11. First aluminum can pull-tab, Iron City Brewery, 1962
12. First ambulance staffed by paramedics, Freedom House Ambulance Service, 1967
13. First night World Series game, game 4 between the Pittsburgh Pirates and Baltimore Orioles, 1971
14. First Mr. Yuk Sticker, created at the Children's Hospital, 1971
15. First internet emoticon, the smiley :-), created by Scott Fahlman at Carnegie Mellon University, 1982

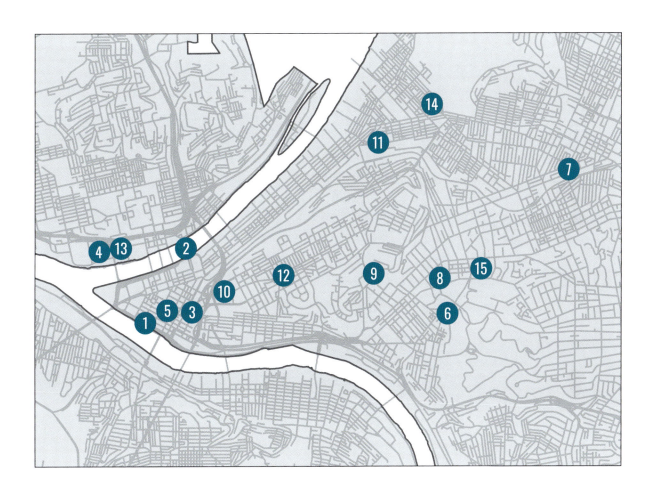

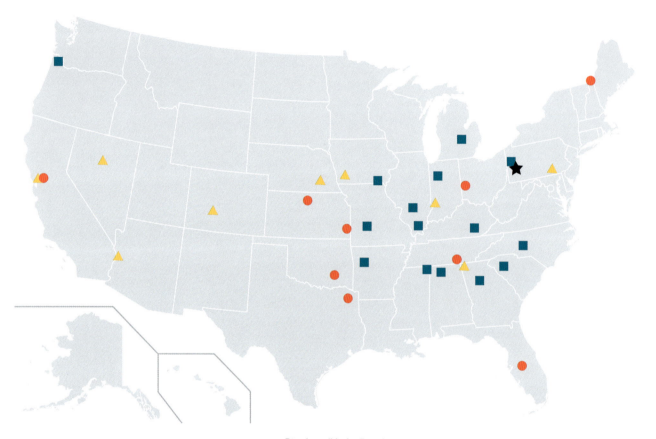

Pittsburgh With an *h*

The name "Pittsburgh," with a silent *h*, was bestowed on the settlement at the forks of the Ohio River by General John Forbes in 1758, in honor of British Prime Minister William Pitt the Elder. Because Forbes was Scottish, historians surmise that he intended the name to be pronounced "Pitts-burrah," like Edinburgh in Scotland, but that pronunciation never caught on.

Because Pittsburgh was a major city in the early American republic and a common jumping-off point for settlers moving west, thirty-three other towns across the country ended up being named after the Pennsylvania city.

In 1891, the US Board on Geographic Names decided to standardize place-names around the country, decreeing that all "-burgh" names should drop their silent *h*. This affected the original Pittsburg(h) as well as its namesakes. For the next two decades, the use of the h was mixed in the city, with some institutions—such as the city government and the Pirates baseball team—dropping it, while many other businesses and citizens stuck to the traditional spelling.

In 1911, Pennsylvania senator George T. Oliver convinced the board to make an exception for Pittsburgh, Pennsylvania. On June 19 of that year, the board approved the change, and Pittsburgh officially got its *h* back. The anniversary is celebrated by some Pittsburghers today as "H" Day. The *h* features in the popular abbreviation "PGH" and its derivatives, such as the Pittsburgh Water and Sewer Authority's "PGH2O" logo.

Yinz N'At

Linguists say that Pittsburghese—officially, Western Pennsylvania English—is the result of the area's unique migration history, with important early influences from the speech of Scots-Irish, Pennsylvania Dutch, Polish, Ukrainian, and Croatian immigrants. Mass media has caused the decline of regional accents all around the country, and today the strongest Pittsburghese accents can be found among older working-class people. But the dialect has also become a source of city pride, so don't count it out just yet.

Pittsburghers call ourselves "Yinzers," after the city's unique second person plural pronoun "yinz" (a contraction of "you ones"). Other notable bits of Pittsburghese lingo include "n'at" ("and that," and so on), "redd up" (tidy), "dippy" eggs (with runny yolks), "jagger" (thorn), "jagoff" (jerk or stupid person), and "nebby" (nosy).

Yinzers also pronounce things distinctively—for example, the "ow" sound becomes "ah," so we go dahntahn instead of downtown. The so-called "cot-caught merger" means that those two words sound exactly the same in Pittsburghese, as do "dawn" and "Don." The "ee" sound sometimes turns into a short "i," so we root for the Stillers, swim in the crick, and get groceries at Giant Iggle. Even the city's name is affected: a real Yinzer will tell you they hail from "Pixburgh."

This map shows the locations of geotagged posts that used the word "yinz" on Twitter (now X) between August 2022 and February 2023. The concentration in the Pittsburgh area is obvious, and even the tweets posted further afield are usually by Pittsburghers or referring to the city, showing how this one little word has become a marker of Pittsburgh identity and pride. The inset map shows the locations of a number of local businesses with "yinz" in their name, including the Yinz Coffee chain, Yinz Comfy Heating and Cooling, and Yinz Bounce N'At party rentals.

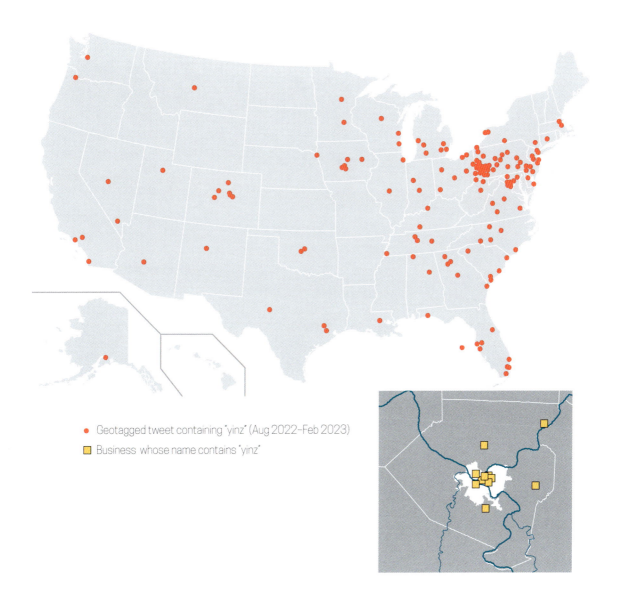

- Geotagged tweet containing "yinz" (Aug 2022–Feb 2023)
- Business whose name contains "yinz"

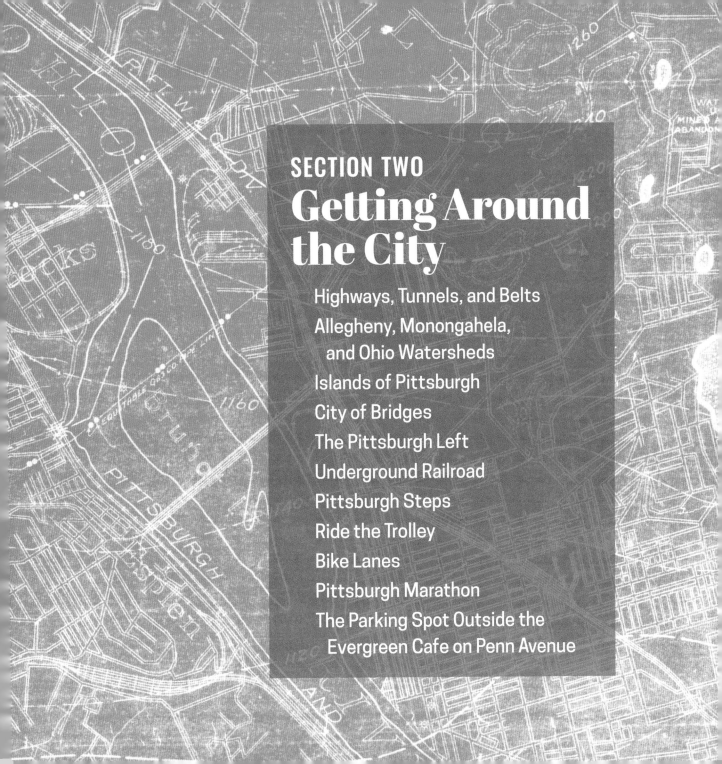

SECTION TWO
Getting Around the City

Highways, Tunnels, and Belts
Allegheny, Monongahela, and Ohio Watersheds
Islands of Pittsburgh
City of Bridges
The Pittsburgh Left
Underground Railroad
Pittsburgh Steps
Ride the Trolley
Bike Lanes
Pittsburgh Marathon
The Parking Spot Outside the Evergreen Cafe on Penn Avenue

Highways, Tunnels, and Belts

Unlike many cities its size, Pittsburgh does not have a dedicated ring road or beltway circling the metropolitan area. A partial ring is formed by Interstate 79 to the west, the Pennsylvania Turnpike (I-76) to the northeast, and the partially completed Southern Beltway (PA 576). Instead of a separate beltway road, Pittsburgh has a unique system of color-coded "belts," made up of existing streets and highways. This system was introduced in the 1950s as an aid to motorists, helping them avoid downtown congestion. An additional belt circling the immediate downtown area was created in the 1990s. The system does still work, even in the age of GPS—shortly after my move back to the city, someone gave me directions by simply telling me to follow the Yellow Belt.

The primary routes into the city are the Parkway—Interstates 376 and 279—or the three river roads: Ohio River Boulevard (PA 65) on the north side of the Ohio; the Allegheny Valley Expressway (PA 28) on the north side of the Allegheny; or, to the south, Saw Mill Run Boulevard (PA 51) along Saw Mill Run.

Because of its hilly terrain, Pittsburgh boasts three major traffic tunnels. The Fort Pitt Tunnel carries the Parkway West through Mount Washington, while the Liberty Tunnel serves the same purpose for Route 51. The Squirrel Hill Tunnel takes the Parkway East through—no surprise—Squirrel Hill. As any Pittsburgher will be quick to tell you, the speed limit doesn't change in the tunnels, so there is no need to slow down!

Allegheny, Monongahela, and Ohio Watersheds

Pittsburgh's three rivers have always been the most important routes in and out of the city. The Allegheny and Monongahela approach from opposite directions, draining watersheds with different geology and different weather, meaning the two rivers' waters may be distinctive in color, running side by side without mixing for miles down the Ohio. The Ohio winds across the country until it joins the Mississippi at Cairo, Illinois; from there, the water (and anything dumped into it) eventually reaches the Gulf of Mexico near New Orleans.

The Allegheny begins near Coudersport in north-central Pennsylvania. It loops northward into New York state before crossing the state line again, where it forms Kinzua Lake. The lake is formed by Kinzua Dam, constructed in 1965 to protect Pittsburgh from flooding. The creation of the dam drowned the last remaining lands of the Seneca Nation in Pennsylvania, forcing residents to move to the Seneca reservation in western New York. The river runs for 325 miles, draining a basin of 11,580 square miles.

The Monongahela begins at the confluence of the West Fork and Tygart rivers in Fairmont, West Virginia. From there, it runs 130 miles north to Pittsburgh. (Measured from the source of the Tygart River, the distance is 165 miles.) The Monongahela basin, 7,340 square miles in all, spans parts of Pennsylvania, West Virginia, and Maryland.

The Ohio, the tenth-longest river in the US, runs 981 miles from Pittsburgh to Cairo, and it drains a basin of 189,422 square miles (which includes the basins of the Allegheny and Monongahela). Notable tributaries within Pennsylvania include Chartiers Creek on the western border of Pittsburgh, and the Beaver River, which is twenty-four miles downstream. The Little Beaver River—which flows past the site of the 2022 train derailment and chemical spill in East Palestine, Ohio—enters near the Ohio/Pennsylvania border. The US Geological Survey defines the Upper Ohio-Beaver basin as a 17,236-square-mile area that drains into the river between the Point and Fish Creek, the last tributary carrying water from Pennsylvania to the Ohio.

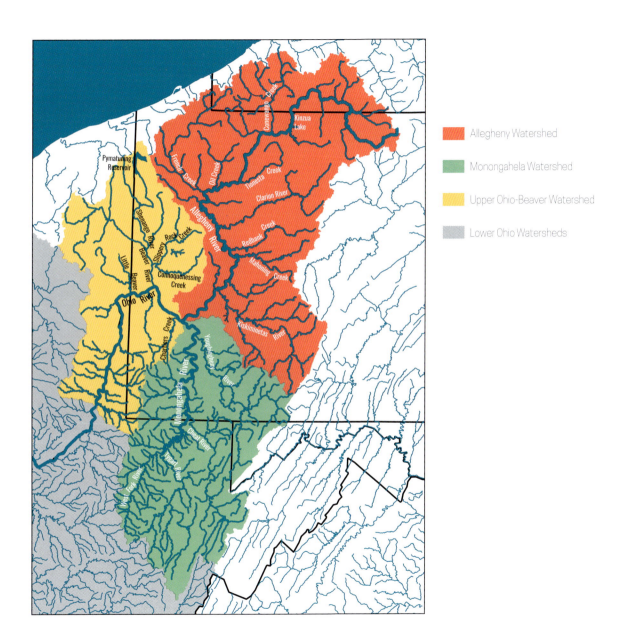

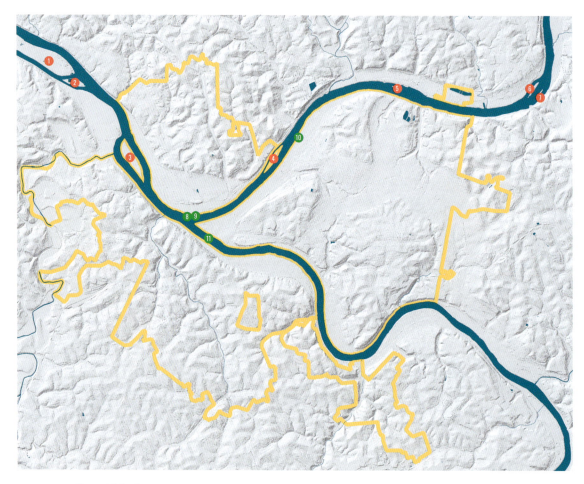

Current Islands
1. Neville Island
2. Davis Island
3. Brunot Island
4. Herr's Island/Washington's Landing
5. Sixmile Island
6. Sycamore Island
7. Ninemile Island

Former Islands
8. Smoky/Killbuck Island
9. Nelson's/Low Island
10. Wainwright's Island
11. Sand Bar/Buckwheat Island

☐ City boundary

Islands of Pittsburgh

With three wide rivers defining its geography, islands are an integral part of Pittsburgh's landscape. Today's rivers have fewer islands than they did in the past, as a number of islands have been lost to shore development over the years.

Several islands were incorporated into the Allegheny River's shores through filling the channel to facilitate industrial development in the late nineteenth century. These include ones known as Smoky or Killbuck Island and Nelson's Island on the north shore near the Point, and Wainwright's Island in Lawrenceville.

The most notable remaining island in the Allegheny is Herr's Island, also known as Washington's Landing. Though it is officially part of the Troy Hill neighborhood, the island's fancy homes and marinas give it a distinct upper-class character in contrast to the rough blue-collar mainland. Continuing upriver, the uninhabited Sixmile Island, managed by the Allegheny Land Trust, lies just on the Sharpsburg side of the city line. Sycamore Island (a great spot to hunt for mushrooms) and Ninemile Island form parts of Blawnox and Penn Hills, respectively.

In the Ohio, the most prominent island is Brunot Island (pronounced with a silent t), home to an oil- and gas-fired electric power plant whose employees commute to work along a pedestrian bridge that extends from the north shore. Farther downstream, one comes to Davis Island (owned by the West View Water Authority) and Neville Island (a township home to 1,100 people).

Today, the Monongahela is entirely island free for the 130 miles it traverses from its source in Fairmont, West Virginia, to the Point in Pittsburgh. But a 1795 map shows a "sand bar" in the center of the Mon, stretching nearly the full length of downtown, with a note reading "Buckwheat grown in 1795." This island was likely destroyed by erosion soon after, as it appears on no later maps.

City of Bridges

Pittsburgh is famous for its bridges, though despite rumors to the contrary, the city does not have the most bridges of any city in the world, or even in the US. The exact count depends on your definition of a "bridge": the US Department of Transportation says that, in the city, there are 297 raised public road spans of twenty feet or more; in his 2006 book, *The Bridges of Pittsburgh*, Bob Regan included footbridges and rail bridges to arrive at his count of 446; and Todd Wilson of Engineering Pittsburgh counted over seven hundred by including any sort of elevated crossing, like walkways between buildings. However you count them, there's no question that bridges are significant in defining Pittsburgh's landscape.

The city's bridges range from the ordinary to the iconic. Among the most celebrated are the "Three Sisters," identical self-anchored suspension bridges connecting the North Shore to Downtown, named for Pittsburgh notables Roberto Clemente, Andy Warhol, and Rachel Carson. Among the most frustrating are the Fort Duquesne and Fort Pitt bridges, where drivers moving between the Parkway North, Parkway West, Parkway East, and various local exits have only a few hundred feet to merge and swap lanes lest they get whisked off in an unintended direction. The Smithfield Street Bridge, over the Monongahela, is the city's longest at 1,184 feet, and it sits on the site of the city's first bridge (the 1818 Monongahela Bridge).

At times, Pittsburgh bridges have attracted unfavorable attention. The Greenfield Bridge notoriously began to crumble and drop chunks of concrete onto the Parkway, necessitating the construction of an "under bridge" to protect the roadway between 2003 and 2015, when the main bridge was replaced. More recently, in January 2022, the Fern Hollow Bridge, which carries Forbes Avenue over a ravine in Frick Park, collapsed, sparking a national conversation about infrastructure maintenance. The bridge was replaced and reopened by the end of the year. A 2024 study found that the number of Pennsylvania bridges rated in "poor" condition has decreased from 24% to 13% over the past decade, but that progress will stall if continued funding for repair is not found.

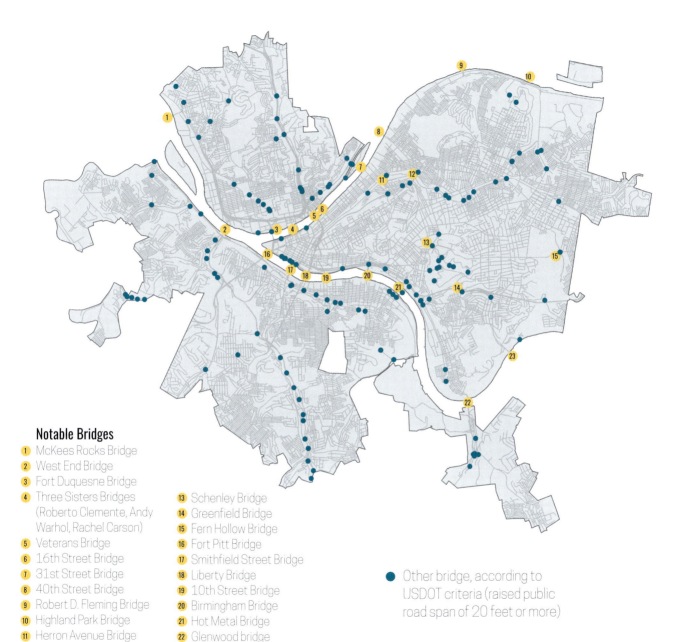

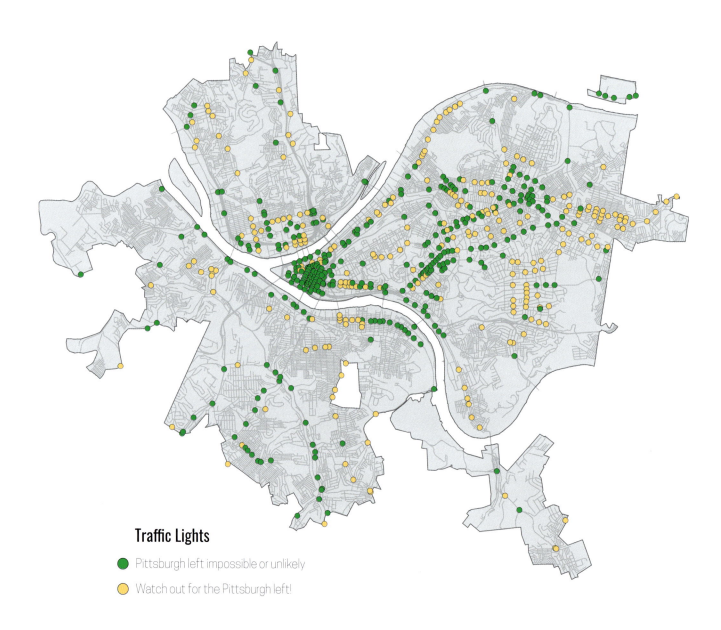

The Pittsburgh Left

Waiting to turn left at a Pittsburgh traffic light, you may find the first driver in the oncoming lane hesitating and flashing their lights to let you know you can turn in front of them. You've just been offered the chance to make a "Pittsburgh left."

At its best, the Pittsburgh left helps keep traffic from jamming up in spots where there's no dedicated left turn lane. Nevertheless, it is important to note that according to the Commonwealth of Pennsylvania, the Pittsburgh left is illegal. The state driver's manual states bluntly, "Drivers turning left must yield to oncoming vehicles going straight ahead." A crash can occur if the turning driver insists on going while the oncoming driver fails to yield in time. Nevertheless, the practice is so common that anyone driving through the city would be well advised to be prepared to have other drivers wait for you to make a left, or to try to make a left in front of you at a light.

This map shows places in the city where you might encounter someone making a Pittsburgh left. I counted an intersection as a potential Pittsburgh left location if it (1) has a traffic light, (2) has only one lane of traffic in at least one direction, and (3) has no dedicated turn lane or arrow for traffic making a left turn across the oncoming lane in that direction.

Underground Railroad

The end of slavery began in Pennsylvania in 1780, with the passage of the Gradual Abolition Act. But just fifty miles south of the city, the Mason-Dixon Line separated Pennsylvania from Maryland and Virginia, where slavery remained legal. This led Pittsburgh to become a hub of the Underground Railroad, the system of routes and safe houses that enabled enslaved people to escape to freedom in northern states and Canada. The Monongahela River and the road from the city of Washington, Pennsylvania, were important pathways that brought formerly enslaved people into Pittsburgh. Once here, they were sheltered by Black and white abolitionists.

Many of Pittsburgh's free Black residents lived in Arthursville and Little Haiti, in the Lower Hill area. These neighborhoods have since been demolished to construct Interstate 576 and the hockey stadiums. Residents including Rev. Lewis Woodson, Samuel Bruce, George Gardner, and Bishop Benjamin Tanner sheltered people fleeing slavery. The Wylie Avenue African Methodist Episcopal Church was home to one of several "vigilance committees" that protected people escaping from slavery.

Downtown, people escaping slavery could get a haircut and change of clothes at the barbershop and bathhouse run by Black abolitionist John Vashon, who hosted the first meeting of the Pittsburgh Anti-Slavery Society in 1833. He also cofounded the Pittsburgh African Education Society. One of Vashon's students was physician Martin R. Delany. In 1843, he launched the abolitionist newspaper the *Mystery*, the first Black-owned newspaper west of the Allegheny Mountains. His home in Downtown Pittsburgh was also used as a safe house along the Underground Railroad. Another Black businessman offering shelter was John C. Peck, owner of an oyster house and a deacon at the Wylie AME church.

Among Pittsburgh's white abolitionists were businessman Charles Avery, hotelier James McDonald Crossan, and lawyer Thomas Bigham. Avery set up the Allegheny Institute and Mission Church in what was then Allegheny City to provide elementary and secondary education to free Black students regardless of gender. Crossan's Monongahela House hotel employed many free Black people, including noted abolitionists Thomas and Frances Scroggins Brown. The staff were

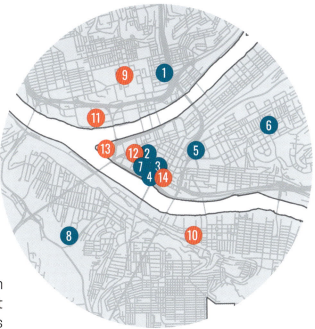

known to use a variety of tactics, including violence, to disrupt the efforts of visiting slave owners seeking to recapture fugitives, or to separate enslaved guests from their masters. Secluded among the trees, and commanding a view of the surrounding land, Bigham's house on Mount Washington was well placed to protect those escaping slavery. A quilt or lantern would be placed on the balcony to signal that the house was safe. A secret extra floor was concealed above the building's upper ceilings.

The end of slavery is celebrated on Juneteenth. This holiday originated in Texas (where enslaved people first learned of their freedom on June 19, 1865) and was first brought to Pittsburgh in 1993 by radio station WAMO. Juneteenth gained wider recognition in Pittsburgh after the 2020 racial justice protests, and in 2024 (after some controversy) two separate celebrations were funded—the Western Pennsylvania Juneteenth organized since 2013 by William "B" Marshall and the Poise Foundation, and a new Juneteenth FusionFest event organized by the city itself through Bounce Marketing.

Underground Railroad Stations
1. Allegheny Institute and Mission Church
2. John Vashon Barbershop and Bathhouse
3. Martin R. Delany House
4. Monongahela House Hotel
5. Arthursville and Little Haiti
6. Wylie AME Church
7. Peck Original Oyster House
8. Bigham House

Juneteenth
9. WAMO Juneteenth (1993)
10. WAMO Juneteenth (1994-c. 2000)
11. Western PA Juneteenth (2013)
12. Western PA Juneteenth (2014-17)
13. Western PA Juneteenth (2018-24)
14. Juneteenth FusionFest (2024)

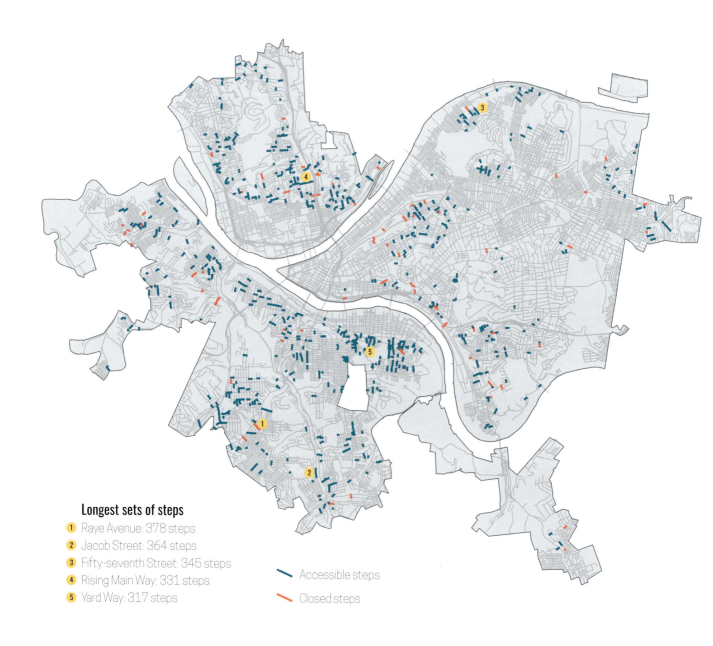

Pittsburgh Steps

Pittsburgh has eight hundred sets of public steps, comprising more than forty-five thousand individual steps. This is more sets of steps than any other city in the US. The steps allow pedestrians to make connections along hillsides that are far too steep for a car-navigable road or conventional sidewalk; they were essential for workers trying to get from hillside immigrant communities to industrial workplaces down near the rivers.

Pittsburgh's steps range from short flights of a few steps to massive ascents of hundreds of steps on staircases that stand dozens of feet above the ground below. Most of the steps are made of concrete, though some are wood or stone. They are all public right-of-ways, and in some cases, they even bear street names. A number of houses in the city are accessible only by city steps.

Unfortunately, maintenance on the steps has not kept up with wear and tear, and a little under 7 percent of them have been officially closed due to safety concerns. The city was recently awarded $7 million in federal funding to repair some of the steps, but with repair costs running between half a million and 1.5 million dollars per staircase, the city faces a tough challenge in making equitable choices about which steps to prioritize for repair.

Ride the Trolley

Seventy-five years ago, Pittsburghers were spoiled for choice in their rail options. Dozens of trolley lines ran through the city's major neighborhoods. But as cars became the dominant mode of travel and public transit shifted to the use of buses in the 1960s, the lines began to close.

Today, if you want to get around Pittsburgh by rail, your options are rather limited. We have a light rail system, known as the T, with two lines that begin at the sports stadiums on the North Shore, cross through Downtown, and wind their way through the South Hills. There are also three busways, which use buses on dedicated routes with stations, kind of like a train. A Bus Rapid Transit line, consisting of dedicated bus lanes, is expected to open between Downtown and Oakland in 2025. Amtrak also maintains a station Downtown, with daily trains to Chicago, DC, and Philadelphia.

A distinctive component of Pittsburgh's rail infrastructure is its funicular railways, known locally as inclines. These are steep rail tracks going up the side of a hill, with a car that sits on an inclined plane to keep it level. Some twenty-four inclines once operated in the city, most of them built in the middle or end of the nineteenth century. Two inclines remain in operation today, both on the north face of Mount Washington.

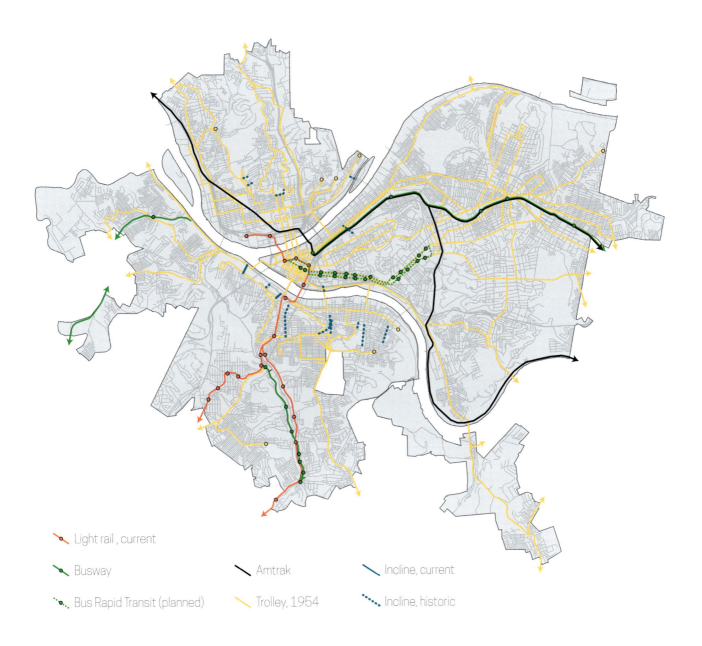

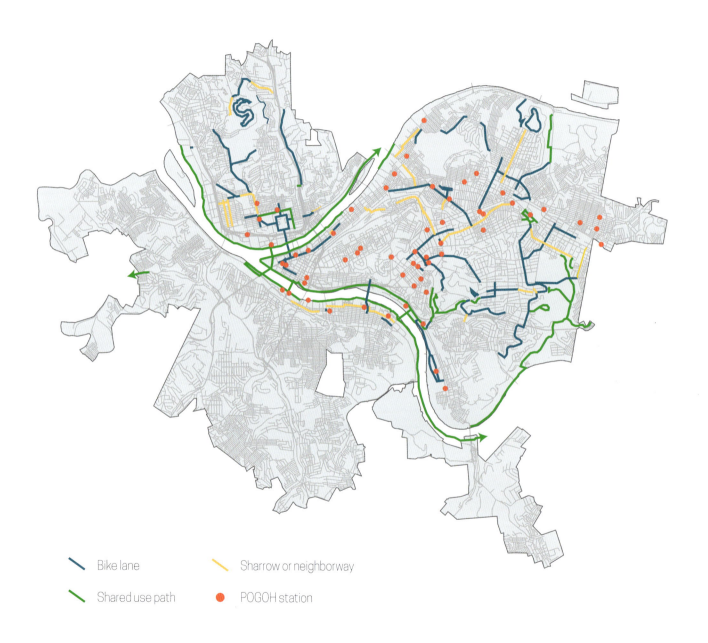

Bike Lanes

In the mid-1990s, Pittsburgh was rated the "worst city in the US to ride a bike." A lack of bike infrastructure along the city's narrow, hilly streets put cyclists at high risk of dangerous encounters with cars. A lot has changed since then, both through efforts by the city itself and by Bike PGH, a local nonprofit. Pittsburgh currently has 102 miles of bike infrastructure. This consists of bike lanes, shared pedestrian and bike paths, sharrows (traffic lanes designated to be shared with bicycles), and neighborways (streets with traffic-calming measures to make them safer for cyclists). Further improvements are planned that will link the city's existing infrastructure together. Just under 15 percent of commuters in the city use a bike at least some of the time.

The city's bike share program, POGOH, currently has sixty stations with six hundred bikes to rent, including both e-assist and pedal bikes. In 2023, more than 211,000 trips were made using POGOH bikes.

Pittsburgh Marathon

On the first Sunday in May, Pittsburgh hosts a marathon. The race begins and ends downtown, and over its 26.2-mile course, it crosses all three rivers and loops through many of the city's most notable locations, with 278 feet of elevation change. Runners are treated to views of the sports stadiums on the North Side; hip bars and restaurants along Carson Street; the Pitt and CMU campuses; the shopping district on Walnut Street; the stately homes of Highland Park; and the famous markets of the Strip District. In 2024, some forty-two thousand people participated in one of the events during race weekend, which include the marathon, half marathon, kids' marathon, and 5K; some three hundred thousand spectators came out to watch them run. Participants ranged in age from 5 months to 87 years old. Organizers prepared with 20,000 gallons of water and 33,600 bananas. The winners of the men's and women's races each took home a $7,000 prize.

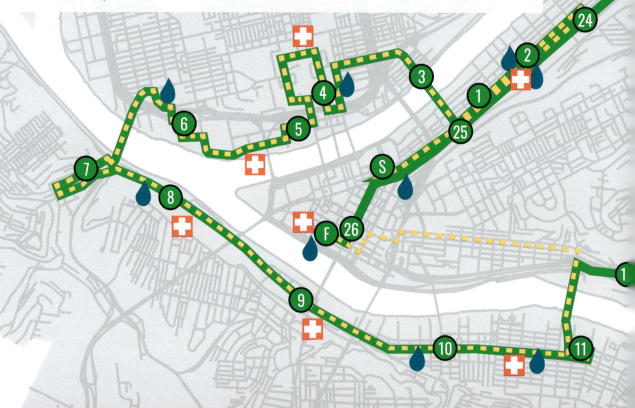

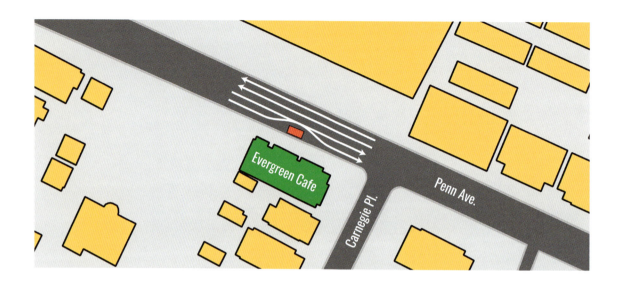

1. Center for Creative Reuse
2. Construction Junction
3. Evergreen Cafe
4. East End Food Co-op
5. Reformed Theological Seminary
6. Frick Art and Historical Center
7. Frick Park
8. Homewood Cemetery

The Parking Spot Outside the Evergreen Cafe on Penn Avenue

For most of its length from East Liberty to Wilkinsburg, Penn Avenue is a four-lane thoroughfare. But outside the Evergreen Cafe at 7330 Penn Avenue, a single parking space occupies the right lane on the eastbound side, forcing drivers to merge into the left lane.

The owner of the Evergreen, Phil Bacharach, has kept his vehicle parked in that space regularly since the 1970s. Frustrated drivers have tried calling the police, and some have even hit Bacharach's car, but the spot remained legal parking. Readers of City Paper voted him the city's "Best Jagoff" in 2023, a designation Bacharach wears with pride—his car even sports a "BEST JAG" license plate.

Penn Avenue is a major route for commuters coming in and out of the eastern part of the city, connecting to Route 8 (and thence to the Parkway East) not far from the Evergreen. It's also heavily used by people visiting the various amenities in the neighborhood, such as the East End Food Co-op, Construction Junction and the Center for Creative Reuse (where unwanted building materials and craft supplies, respectively, are re-sold to keep them out of the landfill), the Frick Art and Historical Center, Frick Park, and Homewood Cemetery.

In 2023, the collapse of the Fern Hollow Bridge diverted traffic from Forbes Avenue onto this stretch of Penn, and the conflict really heated up. In January 2024, as a result of a petition filed with the city, the location was changed to a thirty-minute loading zone. The petition, posted on change.org by Regent Square resident Jennifer Makovics and signed by nearly eight hundred users, calls on the city to "prioritize safety and traffic flow over a single person's convenience." A competing change.org petition created by Bacharach's daughter, with over seven hundred signatures, calls Makovics's effort a "smear campaign against a small business owner" and points out that the spot is also used by Evergreen customers and a local food truck. Bacharach has vowed to fight back, complaining that the city did not properly notify him or explain the reason for the change. As of the writing of this chapter in June 2024, he continues to occupy the space for "loading" as often as possible.

SECTION THREE
Communities and Neighborhoods

A City of Neighborhoods
Mount Oliver
Gentrification
Old Houses and Lead Paint
Children and Seniors
Race and Ethnicity
Tree Cover
Supermarkets and Food Deserts
Chinatown
Immigrant Population
LGBTQ+Burgh
Jewish Pittsburgh
Catholic Pittsburgh
Muslim Pittsburgh
Mexican War Streets
Blue Dot in a Red Sea
To Outer Space

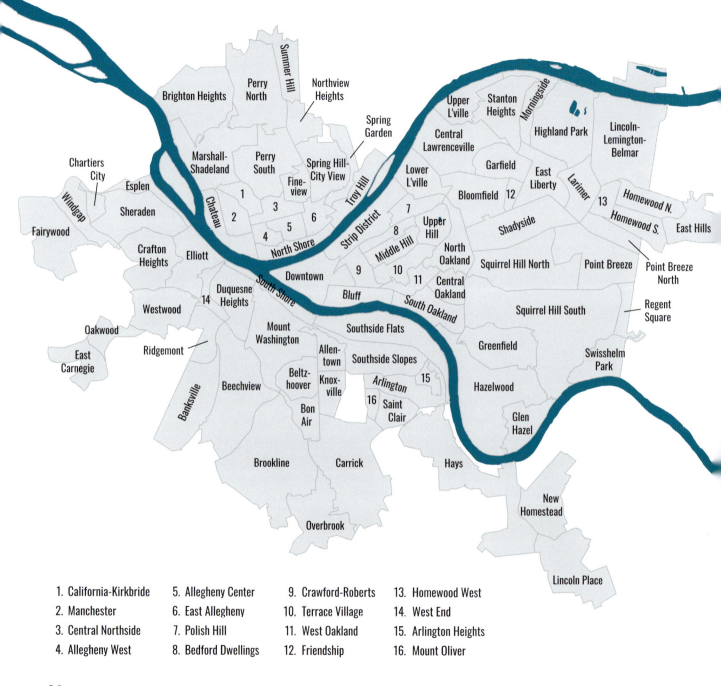

A City of Neighborhoods

Pittsburgh is made up of ninety unique neighborhoods, each with its own distinct character. South Oakland is where college students seek out cheap apartments, while Shadyside boasts million-dollar mansions. The North Shore is for sports fans, Southside Flats is for partying on a Friday or Saturday, and Mount Washington has the most romantic views. Many are known for particular ethnic communities that have settled there and left their mark on the architecture and businesses—Latinos in Beechview, Italians in Bloomfield, Jews in Squirrel Hill, and (unsurprisingly) Poles in Polish Hill.

Pittsburgh's neighborhoods are clearly defined by the city. In many cases, the borders follow the city's topography, where steep hillsides create gaps in the street network and thereby foster a sense of separation between residents on either side. This is not to say there isn't still controversy sometimes, particularly around naming. The area known officially as East Allegheny, for example, has been rebranded Deutschtown by economic boosters—apparently "Deutschtown Music Festival" sounds cooler than "East Allegheny Music Festival."

As of the 2020 Census, the largest neighborhoods by population were Squirrel Hill South and Shadyside, with 15,317 residents each, and the smallest was Chateau, which boasts nineteen. Squirrel Hill South, at 2.677 square miles, is also the largest neighborhood by land area, while the smallest is Mount Oliver, at 0.103 square miles (note that this is the Mount Oliver neighborhood, which sits just east of the 0.34-square-mile borough of Mount Oliver).

Mount Oliver

Mount Oliver, in the city's southern hills, is the name of a neighborhood of Pittsburgh, and it's also an independent borough entirely surrounded by the city of Pittsburgh. The area was first settled in 1769 by John Ormsby, an officer under the command of General Forbes during the French and Indian War, who had a son named Oliver. The settlement was initially part of St. Clair Township, but in 1892, a group of residents petitioned to incorporate as an independent borough in order to resist Pittsburgh's expansion efforts. The unincorporated portions of Mount Oliver did get annexed, becoming the Pittsburgh neighborhood. In 1927, Pittsburgh failed to force the borough's annexation. Residents remain fiercely proud of their independence and value how quickly their concerns can be addressed by their local government. The borough has its own mayor and seven-member council and maintains a separate police and fire department. At the same time, the two municipalities are deeply intertwined, with little more than a change in street sign color (blue in the city, green in the borough) to tell you you've crossed the border. Mount Oliver is part of the Pittsburgh Public Schools District, and residents have to pay school tax to the city while the remainder of their local taxes go to the borough.

The main business district of Mount Oliver, along Brownsville Road, includes restaurants, banks, and hair and nail salons. The borough has three churches—nondenominational, Baptist, and Lutheran, the latter of which also has a cemetery where many early German settlers are buried. Recreational facilities include a playground and a sports field.

Mount Oliver had 3,394 residents as of the 2020 Census. The population is 47 percent white and 38 percent Black. Nearly 19 percent of residents reported German ancestry, and there are also sizeable populations of those with Italian, Polish, and Irish roots—a very typical Pittsburgh-area combination.

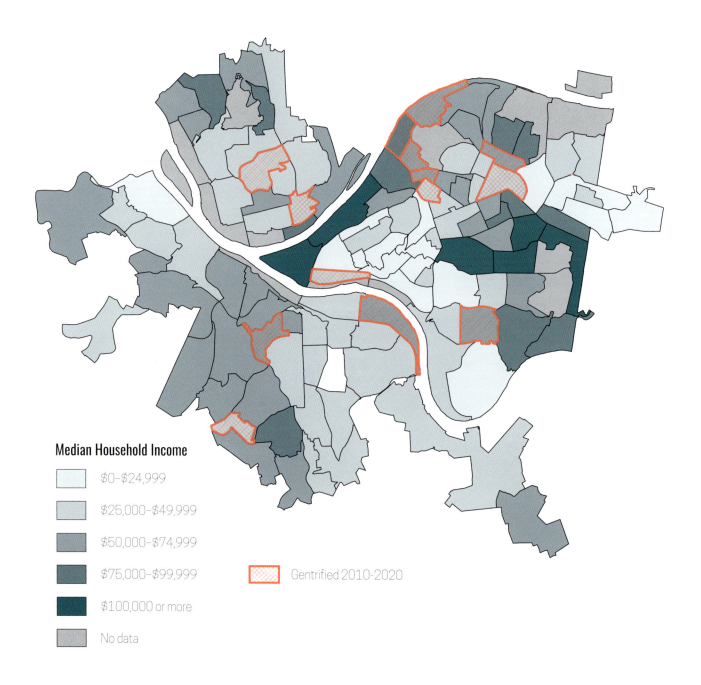

Gentrification

Gentrification occurs when a poorer area of a city is targeted for redevelopment, in the hopes of drawing in a new, wealthier population. This process is often embraced by city leaders because it expands the city's tax base; on the other hand, more often than not, it displaces the area's previous residents, forcing them to areas of the city where they are farther away from jobs, transit options, and cultural amenities that were present in their old neighborhood.

Creating a consistent quantitative measure of gentrification is a tricky proposition. In this map, I've used a simple measure created by the Urban Health Collaborative at Drexel University. A census tract is considered to have been gentrified if it had a below-median home value in 2010 and above-median increases in home values and the percentage of residents with college education between 2010 and 2020. By this measure, gentrification happened in parts of the North Side, Lawrenceville, and East Liberty, as well as other Pittsburgh neighborhoods.

A 2019 study by the National Community Reinvestment Coalition rated Pittsburgh the eighth most gentrified city in the country, so it's no surprise that gentrification is a frequent topic of public debate. Efforts to resist gentrification include the city's land bank—which encourages the redevelopment of vacant lots into affordable housing and community spaces—and cooperative agreements between Penguins team owners and residents of the Hill District, promising benefits to local communities from redevelopment of areas around the hockey rink.

Conflict over gentrification in East Liberty came to a head in 2018 when a property management company demanded the removal of a billboard on which local artist Alicia Wormsley had written "there are Black people in the future." The billboard was atop a building that formerly hosted the Shadow Lounge, a prominent Black bar and music venue that had closed due in part to gentrification of the neighborhood. A spokesperson for the property company called the billboard "offensive and divisive," but many community members viewed the removal as symbolic of the way that gentrification has pushed out the city's Black residents.

Old Houses and Lead Paint

Much of Pittsburgh's housing stock dates back before 1940 (the oldest recorded by the US Census). Some of these older homes are stately mansions that have been maintained, while others are in dire need of repair and present health risks to inhabitants. While some neighborhoods have experienced recent redevelopment—the Central Northside, the Strip District, East Liberty, and the Hill District—in other neighborhoods—including areas in the North Side, Oakland, South Side, and along Frick Park—the city has acted to preserve older housing in certain areas officially designated as historic districts.

Surviving homes of notable Pittsburghers include those of engineer James Andrews (built 1866), astronomer John Brashear (built 1886), mystery writer Mary Roberts Rinehart (built 1890), and steel executive James Scott (built 1900). Clayton, the Point Breeze mansion built by steel magnate Henry Clay Frick, has been converted into a museum, as has the childhood home of playwright August Wilson (built c. 1840). Built in 1790, the John Frew House in Westwood is the only occupied residential property (and one of only five total surviving buildings) that was constructed before the city's incorporation in 1816. Chatham Village in Mount Washington (built 1932-1936) is an iconic example of the Garden City movement, which aimed to provide beautiful working-class housing but was quickly taken over by wealthier residents.

Despite Pittsburgh's reputation for a low cost of living when compared to out-of-control housing markets like New York City or San Francisco, the city's lower income residents can still struggle to find places to live. Habitat for Humanity of Greater Pittsburgh estimates that the region needs an additional fifteen thousand affordable housing units.

When lower income residents do find older housing that is within their budget, it often comes with health risks such as lead paint—potentially present in any home built before 1978. Lead poisoning can lead to cognitive impairments and developmental delays, so the county mandates lead testing of all children. In 2021, 3.87% of children in the city had elevated blood lead levels, as compared to 1.91% countywide. The use of water additives and replacement of lead water lines has greatly reduced the risk from lead in drinking water throughout the city, leaving lead paint as the primary route of exposure.

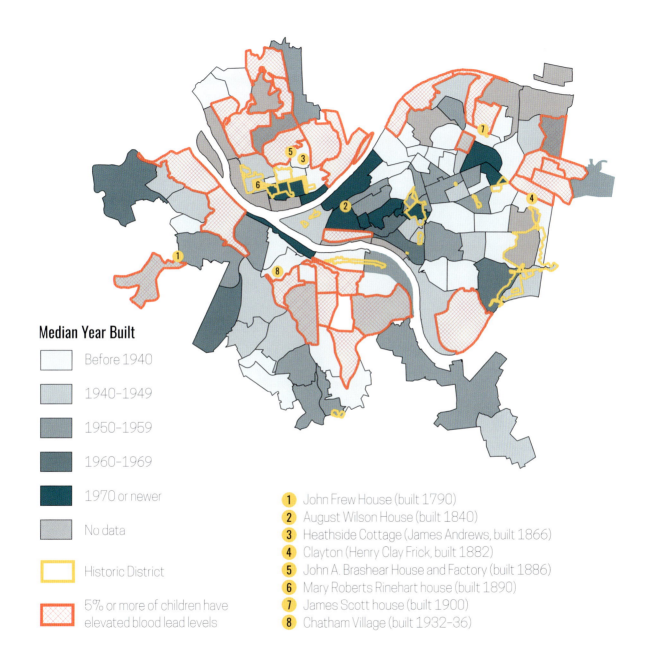

Children and Seniors

Like many Rust Belt cities, Pittsburgh has seen its population growing older over the last few decades. Throughout most of the city, the ratio of children (those under eighteen years old) to elderly (those sixty and over) is below one, meaning that there are more older people than younger. But the story is not the same in all parts of the city. Larger proportions of children can be found in the West End, Hill District, Shadyside, and Homewood, for example. Areas with the fewest children include Fineview, Downtown, and Lawrenceville.

Pittsburgh's youth population has been declining for many years, which is reflected in an 18 percent drop in enrollments at Pittsburgh Public Schools (PPS) from the 2017–18 school year to 2023–24. My employer, Slippery Rock University, gets most of its students from Allegheny and Butler Counties, and the administration has been warning us of a coming "demographic cliff" as these smaller graduating classes move on to college. Shrinking enrollments can lead to shrinking resources and cuts in programs, which in turn push parents (especially wealthier ones) to move out of the district or transfer their kids to private schools, furthering the decline. Yet some schools have begun to turn things around by developing innovative community outreach and partnerships. In the summer of 2024, PPS is holding a series of town hall meetings to discuss plans that may result in consolidation or closure of some schools.

Among the forty most populous counties in the US, Allegheny ranks second—behind retirement destination Palm Beach, Florida—in percentage of the population that is elderly. The high elderly population creates challenges for the county and city to meet increased demand for social and medical services while the tax base declines. Many of Pittsburgh's seniors are people who have "aged in place"; that is, they have continued to live in the same neighborhood (or even the same house) as they got older—more than half of homeowners in the county who are 55 or older have been living in the same house for twenty or more years. Job losses in traditional industries often led younger workers and their families to move away in search of employment, while older people stayed put in the places they knew well.

Race and Ethnicity

Compared to its Rust Belt peers, Pittsburgh lacks racial diversity; the city's population is 62 percent white, as compared to Buffalo at 42 percent, Cleveland at 32 percent, or Detroit at 15 percent. Moreover, Pittsburgh rates high on measures of racial segregation—that is, the extent to which people of different races live in different neighborhoods. An analysis from the University of California at Berkeley rated Pittsburgh the 49th most segregated city in the country (though Pittsburgh is less segregated than Detroit, Cleveland, or Buffalo).

The whitest parts of the city include Lawrenceville, Downtown, the South Side, and many parts of the South Hills neighborhoods. Many of these areas are either home to the city's economic elite, or neighborhoods settled by European immigrant communities that established themselves in the city's industrial heyday. Asian Pittsburghers tend to live in Oakland and Shadyside, many of them having come to study, teach, or research at Pitt or CMU. The city's small Latino population is not as clearly concentrated, though southern neighborhoods like Beechview have notable Latino communities.

Black communities can be found in the Hill District, East Liberty, and Homewood, among others. The current map of Black populations bears a striking resemblance to the areas rated "fourth grade" (out of four) on a 1937 Home Owner's Loan Corporation map that was used to guide racially discriminatory lending. Though such discrimination has been illegal for over half a century, housing and racial justice advocates have documented continuing discrimination that inhibits people of color from moving into traditionally white neighborhoods.

White (non-Hispanic)

Asian (non-Hispanic)

Black (non-Hispanic)

Hispanic or Latino (any race)

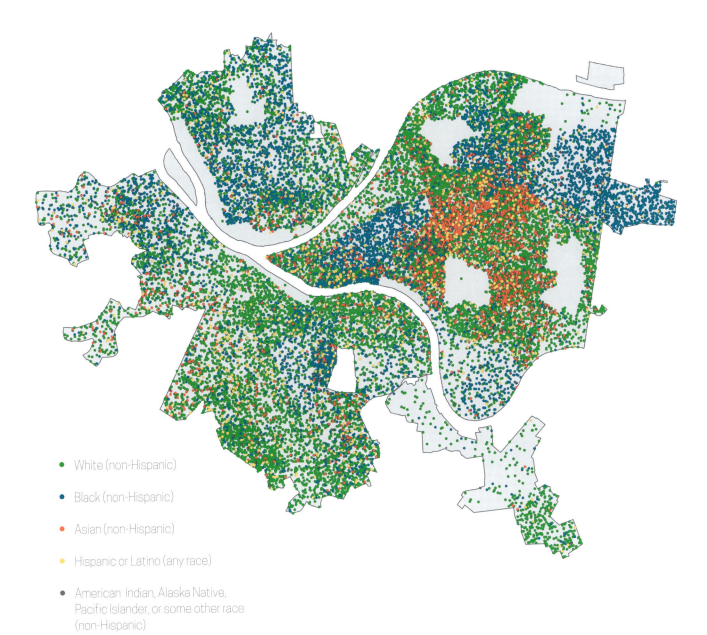

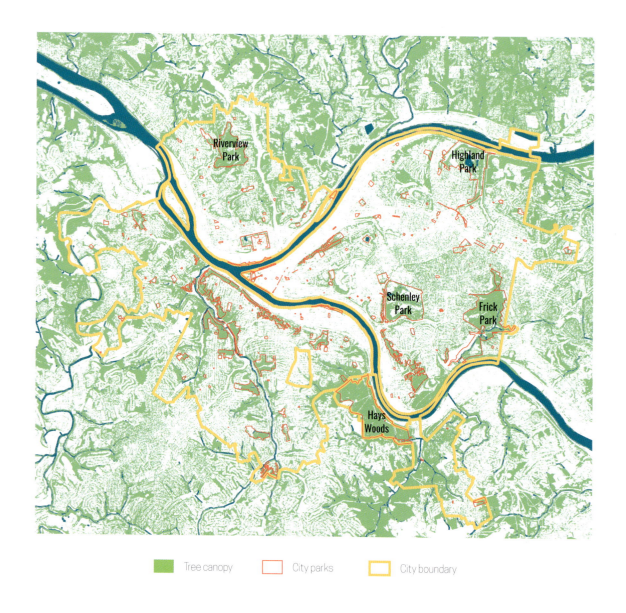

Tree Cover

Five hundred years ago, practically the entire land surface of the Pittsburgh area would have been covered by forest. The landscape was dominated by oak, with maple, beech, and hemlock also present. Light burning by Indigenous people reduced the risk of major wildfires and kept the understory open for easier travel.

Today, urban development has taken a big bite out of the forest. Allegheny County currently has about 52 percent tree cover, while Pittsburgh has 41 percent. This tree cover is also distributed unevenly. The city's major parks—Riverview, Schenley, Highland, Hays Woods, and Frick—boast large expanses of forest, while neighborhoods like Chateau, Bluff, and East Allegheny have less than 15 percent tree cover. Hays Woods is the newest of the city's major parks, acquired in 2021. Though it covers six hundred acres and boasts extensive trails as well as a nesting pair of bald eagles, it remains difficult to access.

Trees provide a variety of benefits, including improving air quality, decreasing summer temperatures, absorbing stormwater, stabilizing soil, encouraging outdoor recreation, and boosting economic activity by making neighborhoods more pleasant to live and shop in. Unsurprisingly, it's often low-income and minority neighborhoods that have less access to urban forest, either because of a lack of trees or because their forested areas are poorly maintained and difficult to access. In recent years, the city has worked with nonprofit Tree Pittsburgh to target programs at planting trees in areas where they're most needed, including Beltzhoover, California-Kirkbride, Homewood, Lawrenceville, Manchester/Chateau, and Hazelwood.

Supermarkets and Food Deserts

In Pittsburgh, as in most cities, grocery stores are not distributed evenly. The Strip District is famous for its many specialty grocery stores and food markets, and areas like Shadyside and the South Side boast ample options as well. On the other hand, some neighborhoods' residents have no solid grocery options within a mile (without crossing a river). In the north, these include Summer Hill, Northview Heights, Spring Garden, and Troy Hill. To the west, Elliott and West View lack close, available grocery stores as well, and residents of Lincoln Place and Hays in the south have to travel more than a mile to get to the closest grocery stores, in Homestead.

Areas with poor access to grocery stores are sometimes called "food deserts" (though some activists prefer the term "food apartheid," which emphasizes that the placement of grocery stores is a deliberate social choice, not a natural occurrence). Living in such an area can lead to poorer nutrition and a higher burden of travel for shopping. Food deserts/apartheid are one of the factors behind food insecurity in Pittsburgh, where an estimated sixty-three thousand residents lack consistent access to adequate food. The opening and closing of grocery stores in underserved neighborhoods is a frequent topic of public concern and petitions to the city. Things are looking up for two long-underserved neighborhoods, as Salem's Market opened in the Hill District in 2024 and a new grocery co-op is planned for Hazelwood the following year.

Of course, the quality of the food and experience at different grocery stores can vary, but that's often a matter of subjective opinion. The Giant Eagle in East Allegheny, for example, is sometimes referred to as the "Dirty Bird" and scores only a 3.5/5 rating on Google (other Giant Eagle stores in and around the city usually get at least 4/5). But I'm willing to stick up for that much-maligned store, having shopped there regularly for many years. As grocery stores go, it's perfectly fine!

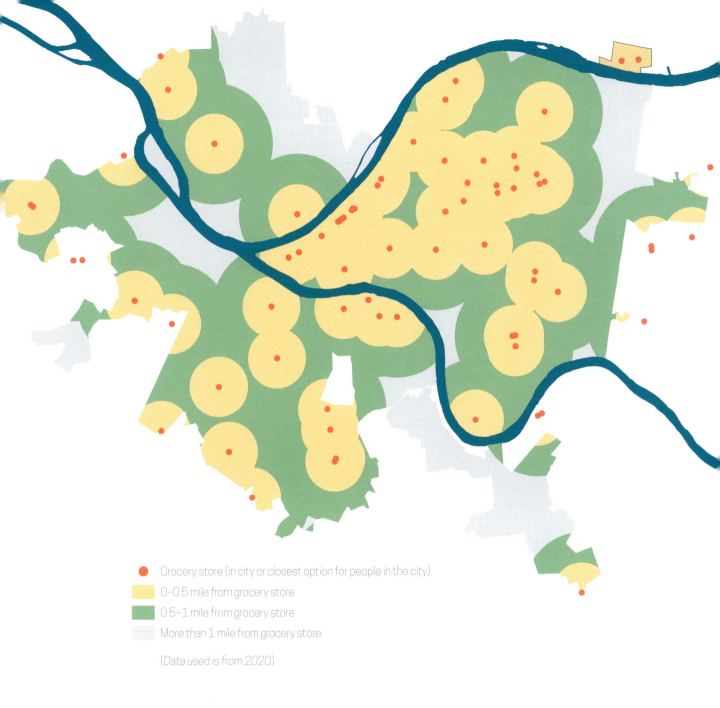

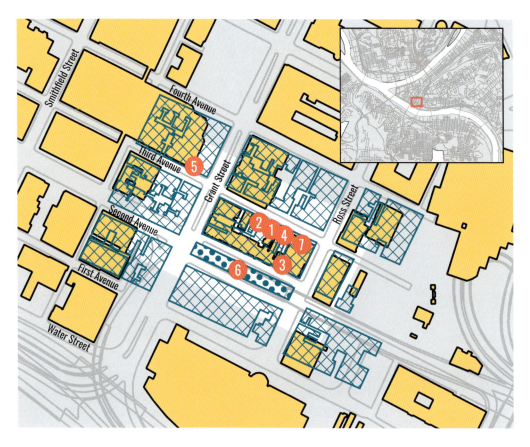

1. Chinatown Inn, formerly Quong Ye Tang grocery store
2. Chong Lee grocery store and Chinese Consolidated Benevolent Association
3. Hip Sing Tong (Cooperating for Success Association)
4. On Leong Tong (Peace and Fraternity Association)
5. Quong Ma Yuen & Co. grocery store
6. Second Avenue Park
7. Sun Wah Shing grocery store

Current buildings
1910 buildings

Chinatown

The first Chinese immigrants arrived in Pittsburgh in the mid-1800s. Most were originally from the same two counties in Guangdong Province; many of them made their way east after the California gold rush petered out. The immigrant community in downtown Pittsburgh grew around Grant Street and Second Street. White Pittsburghers did not extend a warm welcome, and a convention of the American Federation of Labor in the city lent its support to the 1892 Chinese Exclusion Act. Because they were barred from joining white-dominated unions, Chinese Pittsburghers found work in restaurants and laundries (and sometimes as strikebreakers, brought in by factory owners who were eager to exploit ethnic prejudice).

The creation of the Boulevard of the Allies, in the 1920s, destroyed much of the city's historic Chinatown, including many of the buildings inhabited by Chinese Pittsburghers and the Second Avenue Park, which the community used for recreation. Chinese families began to move away from the area, and new arrivals also settled elsewhere in the city. Today, the only remaining piece of the original Chinatown is the Chinatown Inn, located at 520 Third Avenue. Historical records related to Pittsburgh's Chinatown are held at the Heinz History Center.

Today, many Chinese immigrants come to Pittsburgh to work in education or technology. The majority of people of Chinese descent now live in Oakland and Squirrel Hill.

Immigrant Population

For as long as it has existed as a city, Pittsburgh has been a destination for immigrants. Early white settlers came here from England, Scotland, Ireland, and Germany. Later, during the city's industrial heyday, neighborhoods of Italians, Poles, Russians, Serbs, and Croatians grew up around the factories and mills. Currently, immigrants from Asia make up nearly half of Pittsburgh's foreign-born population.

Immigrants make their way to the city for a variety of reasons. Some are drawn here to work in education or technology. Others are refugees, fleeing persecution in their country of origin. For example, the neighborhoods along Route 51 are now home to many Nepali-speaking people who faced ethnic cleansing in Bhutan. The city has also provided refuge to people fleeing the wars in Afghanistan, Syria, Somalia, and Ukraine. Seven European countries—Austria, Czechia, Germany, Italy, the Netherlands, Switzerland, and Slovenia—maintain consulates or honorary consulates in the city, a list more representative of the city's immigrant past than its current immigrant population, though a "mobile" Mexican consulate also visits the city twice a year.

In 2014, the city adopted a "Welcoming Pittsburgh Roadmap" of initiatives aimed at promoting engagement and integration of immigrant communities. For non-citizens needing services from the federal government, the nearest US Citizenship and Immigration Services field office is out in Monroeville. In 2022, the Department of Justice closed the Pittsburgh immigration court, so anyone with a court hearing about their immigration status has to either have a virtual hearing or travel to Philadelphia. A number of local organizations—including AJAPO, Ansar, Bethany Christian Services, the Bhutanese Community Association of Pittsburgh, Casa San Jose, Hello Neighbor, Jewish Family and Community Services, Literacy Pittsburgh, the Pittsburgh Hispanic Development Corporation, and United Somali Bantu of Greater Pittsburgh—help refugees get settled and adapt to life in America.

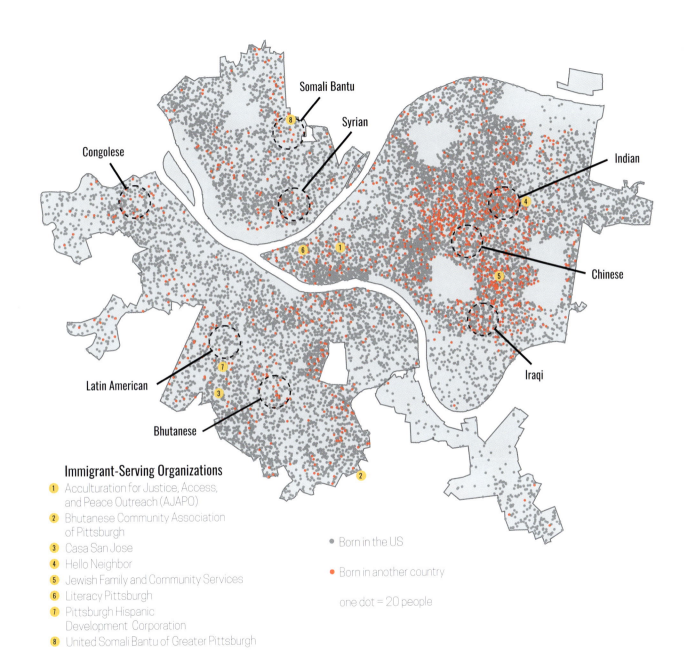

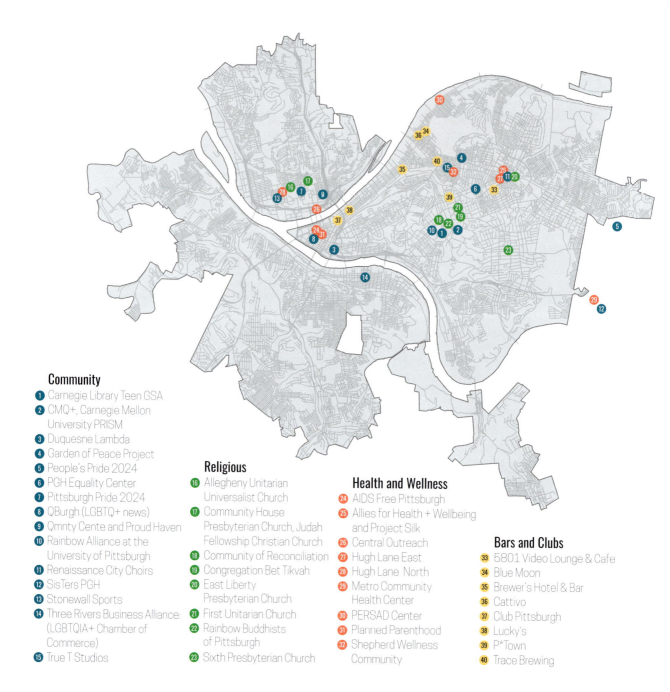

Community
1. Carnegie Library Teen GSA
2. CMQ+, Carnegie Mellon University PRISM
3. Duquesne Lambda
4. Garden of Peace Project
5. People's Pride 2024
6. PGH Equality Center
7. Pittsburgh Pride 2024
8. QBurgh (LGBTQ+ news)
9. Qmnty Cente and Proud Haven
10. Rainbow Alliance at the University of Pittsburgh
11. Renaissance City Choirs
12. SisTers PGH
13. Stonewall Sports
14. Three Rivers Business Alliance: (LGBTQIA+ Chamber of Commerce)
15. True T Studios

Religious
16. Allegheny Unitarian Universalist Church
17. Community House Presbyterian Church, Judah Fellowship Christian Church
18. Community of Reconciliation
19. Congregation Bet Tikvah
20. East Liberty Presbyterian Church
21. First Unitarian Church
22. Rainbow Buddhists of Pittsburgh
23. Sixth Presbyterian Church

Health and Wellness
24. AIDS Free Pittsburgh
25. Allies for Health + Wellbeing and Project Silk
26. Central Outreach
27. Hugh Lane East
28. Hugh Lane North
29. Metro Community Health Center
30. PERSAD Center
31. Planned Parenthood
32. Shepherd Wellness Community

Bars and Clubs
33. 5801 Video Lounge & Cafe
34. Blue Moon
35. Brewer's Hotel & Bar
36. Cattivo
37. Club Pittsburgh
38. Lucky's
39. P*Town
40. Trace Brewing

LGBTQ+Burgh

LGBTQ+ people have been in Pittsburgh for as long as humans have inhabited this land, but the city's first official gay and lesbian institutions, such as bars and bathhouses, began to show up in the 1960s, mostly Downtown or in Oakland. Pittsburgh's first Gay Pride march, in 1973, drew around 150 people, who walked together from Market Square to Schenley Park.

The first case of AIDS in Pittsburgh was documented in 1983. That same year, the University of Pittsburgh Graduate School of Public Health began the Pitt Men's Study, a long-term study of the epidemiology of HIV in men who have sex with men. The study, which is still ongoing, has played a crucial role in understanding and treating the AIDS epidemic.

Pittsburgh Pride went on hiatus during the conservative 1980s, but relaunched in 1991. In the years since, it has grown into a major event for the city, though it has also attracted controversy: the Delta Foundation (which organized the event) was accused of being exclusionary toward trans, nonbinary, and BIPOC individuals; taking sponsorships from companies like EQT, whose values were at odds with many Pride attendees; and mismanagement of funds. Starting in 2017, an alternative event, People's Pride, was organized by SisTers PGH, a Black, trans-led organization. The Delta Foundation was disbanded in 2021, and Pittsburgh Pride was taken over by a coalition made up of TransYOUniting, QBurgh, Proud Haven, and Trans Pride PGH. Today, Pittsburgh Pride and People's Pride co-exist, with the former being held in the North Side and the latter just outside the city limits in Wilkinsburg in 2024.

Beyond Pride events, LGBTQ+ culture has blossomed throughout the area. Many religious organizations now advertise themselves as welcoming, and LGBTQ+ organizations have sprung up at most schools in the city. The city is home to a variety of bars and clubs that cater to the LGBTQ+ community, and there are several dedicated health care organizations.

Jewish Pittsburgh

The earliest documented Jewish residents of Pittsburgh arrived from Germany sometime around 1840. The city's Jewish community initially centered on the Hill District, but by the 1940s, most Jewish institutions had moved to Squirrel Hill, which has become one of the country's preeminent Jewish neighborhoods. The neighborhood, along with parts of Bloomfield, Oakland, and Greenfield, is enclosed by the city's eruv, a string attached to various utility poles and other structures, which creates an enclosed area in which carrying objects outdoors is permitted on the Sabbath. Today, an estimated fifty thousand Jews live in the Pittsburgh area, and the Jewish Federation of Greater Pittsburgh lists thirty-nine congregations in and around the city.

Pittsburgh's Jewish community suffered the country's deadliest act of antisemitic violence on November 4, 2018, when a white supremacist gunman attacked Shabbat services of the Tree of Life/Or L'Simcha, New Light, and Dor Hadash congregations in Squirrel Hill. Eleven people were killed, and six more were wounded. In the tragedy's aftermath, the wider Pittsburgh community rallied to support the victims. Signs and bumper stickers—with the four-pointed stars of the Steelers logo replaced with Stars of David—proclaimed the city was "stronger than hate." The Tree of Life synagogue building is currently being rebuilt to include a museum and memorial to the shooting victims, as well as new worship spaces.

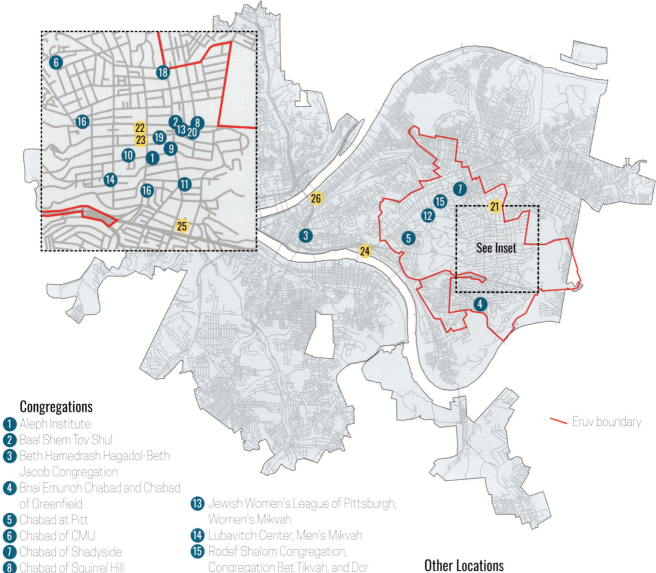

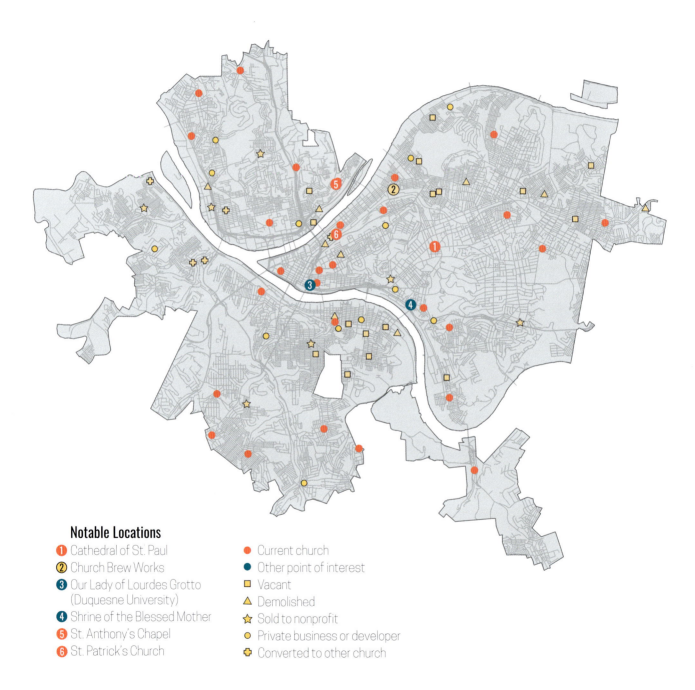

Catholic Pittsburgh

The first Roman Catholic church in the city, St Patrick's, was established in the Strip District in 1808. Catholicism grew rapidly in the late nineteenth century as immigrants from Southern and Eastern Europe flocked here and established their own congregations to preserve particular national traditions. The church buildings they erected were frequently beautiful and majestic, even when parishioners were living in slums. In 1878, Pittsburgh Catholic College, which would later become Duquesne University, was founded in Uptown.

St. Anthony's Chapel in Troy Hill holds more holy relics—more than four thousand—than any church outside the Vatican. Many of these relics, collected in the late nineteenth century by Rev. Suitbert Godfrey Mollinger, were rescued from anti-Catholic campaigns in Germany and Italy. The church's collection includes—purportedly—a piece of the cross on which Jesus was crucified, bones of all twelve apostles, and a tooth of the church's namesake, St. Anthony.

In recent decades, attendance at Catholic services has dropped, and the Diocese of Pittsburgh (which covers most of southwestern Pennsylvania) has been forced to close or consolidate many parishes. The buildings, which remain prominent features of the landscape, have gone on to a variety of uses: as churches for other denominations, as converted apartments and condos, or, in one controversial case, as a brewpub. Thirty-two Catholic churches (out of an original eighty-three) remain active within the city, including the Diocese's Cathedral at St. Paul in Oakland.

In addition to churches, Pittsburgh is also home to several grottoes and other Catholic religious sites. The most famous is the Shrine of the Blessed Mother, also known as Our Lady of the Parkway because it is located high on a hillside overlooking the Parkway East. The shrine was established in 1956, after a local man said he had a vision of the Virgin Mary appearing at that location. Tradition states that a small spring at the site began flowing in response to visitors' prayers.

Muslim Pittsburgh

The first Muslims to come to Pittsburgh were formerly enslaved people, who reverted to the faith of their ancestors from Africa. A thriving Muslim community emerged in the city's Black neighborhoods in the early twentieth century, soon becoming one of the largest Black Muslim communities in the country. In 1932, Pittsburgh's Muslims established al-Masjid al-Awwal, also known as the First Muslim Mosque of Pittsburgh, the first mosque in the US to be founded by native-born Americans. The Nation of Islam gained a following in the city in the mid-twentieth century, establishing its Muhammad Temple No. 22 in Homewood in 1956, and frequently getting into theological and political conflicts with the more orthodox Sunnis of the First Muslim Mosque. After the death of founder Elijah Muhammad, the congregation—now located at Masjid An-Nur in Wilkinsburg—left the Nation and adopted a more universalist form of Islam.

The city's largest mosque (also Sunni), the Islamic Center of Pittsburgh, was founded in Oakland in 1989. It runs a food pantry for people of all faiths as well as numerous Muslim educational programs. The Islamic Center and the city's chapter of the Council on American Islamic Relations are both members of the Pittsburgh Interfaith Impact Network, an organization that brings them together with Jewish, Christian, and Unitarian Universalist groups to promote understanding and advocate for social justice.

Today, Black American Muslims in Pittsburgh have been joined by recent immigrants from Muslim countries such as Pakistan, Somalia, Bosnia, Syria and Iraq. An estimated twenty thousand Muslims live in the greater Pittsburgh area. Outside the city limits, mosques have been built in Etna, Gibsonia, Monroeville, Carnegie, Wilkinsburg, and Moon Township.

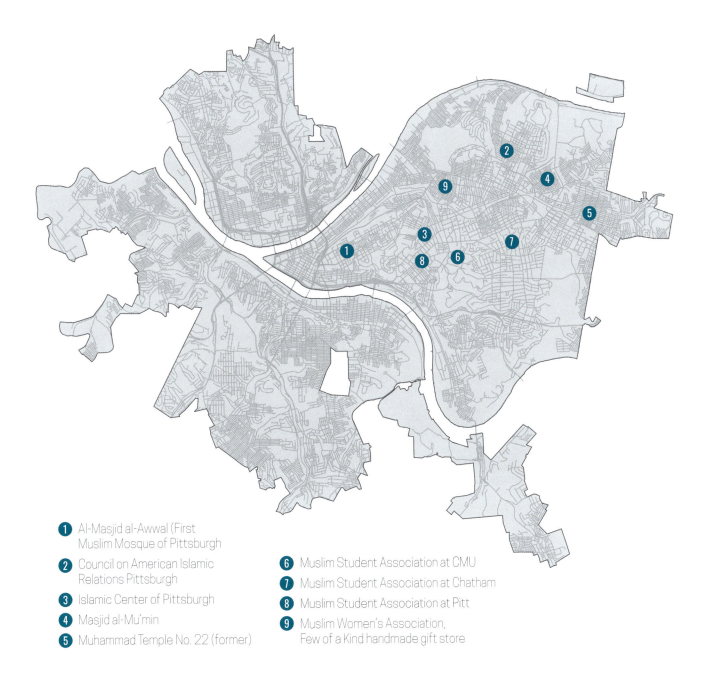

1. Al-Masjid al-Awwal (First Muslim Mosque of Pittsburgh
2. Council on American Islamic Relations Pittsburgh
3. Islamic Center of Pittsburgh
4. Masjid al-Mu'min
5. Muhammad Temple No. 22 (former)
6. Muslim Student Association at CMU
7. Muslim Student Association at Chatham
8. Muslim Student Association at Pitt
9. Muslim Women's Association, Few of a Kind handmade gift store

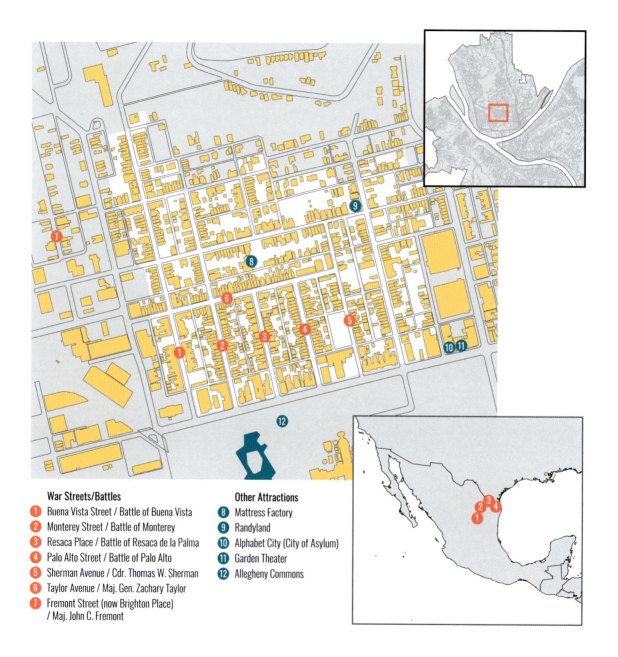

War Streets/Battles
1. Buena Vista Street / Battle of Buena Vista
2. Monterey Street / Battle of Monterey
3. Resaca Place / Battle of Resaca de la Palma
4. Palo Alto Street / Battle of Palo Alto
5. Sherman Avenue / Cdr. Thomas W. Sherman
6. Taylor Avenue / Maj. Gen. Zachary Taylor
7. Fremont Street (now Brighton Place) / Maj. John C. Fremont

Other Attractions
8. Mattress Factory
9. Randyland
10. Alphabet City (City of Asylum)
11. Garden Theater
12. Allegheny Commons

Mexican War Streets

The Mexican War Streets is a historic district in Pittsburgh's Central Northside neighborhood. The area was subdivided for development in 1847 by Gen. William Robinson, and many of the streets were named for places or people associated with the Mexican-American War (1846–1848).

Gen. Robinson, who inherited the land from his father (who had received it as a grant for his service in the Revolutionary War) was a major booster of the Mexican-American War. He had earlier participated in Aaron Burr's failed 1806 attempt to invade and conquer Mexico.

The Mexican War Streets are known for their beautiful, densely built Victorian homes, which were originally inhabited by wealthy residents of what was then Allegheny City. During the twentieth century the area went into economic decline and gained a reputation for crime. But in recent decades, gentrification has taken hold of the neighborhood, renovating and preserving the classic architecture while also driving up housing costs.

Today, the Mexican War Streets are home to a number of attractions. The Mattress Factory museum hosts a variety of art installations, including a lighted work on the roof visible across the neighborhood. Randyland, on the corner of Arch and Jacksonia Streets, is a house and courtyard converted into an elaborate public art display by local artist Randy Gilson. City of Asylum runs the Alphabet City bookshop and performance venue and works to help writers exiled from their countries for their political views. In 2014, City of Asylum sponsored the River of Words public art installation, in which one hundred words in English and Spanish (such as "temperance," "libertad," and "vortex"), selected by Venezuelan artists Israel Centeno, Carolina Arnal, and Gisela Romero, were adopted by neighborhood residents and attached to the outside of their houses. A stroll through the neighborhood will bring up frequent pleasant surprises when you spy a word peeking out of a window or beside a door. The Garden Theater, a prominent former movie theater, is currently undergoing redevelopment. And residents of the War Streets have access to Allegheny Commons Park, just across North Avenue.

As to whether there are any Mexicans in the Mexican War Streets…the US Census lists only 104 Latino or Hispanic people in the whole Central Northside (3% of the population).

Blue Dot in a Red Sea

Political strategist James Carville once quipped that Pennsylvania is Philadelphia and Pittsburgh with Alabama in between. The cities are Democratic strongholds, but the rural areas of the state have more in common, politically, with conservative southern states. Put the two together, and Pennsylvania is a perpetual "swing state," receiving outsized attention and visits from candidates.

In Pittsburgh, Joe Biden beat Donald Trump in the 2020 presidential race by 56.8%. But voters in the suburbs and rural center of the state largely counteracted those votes, and Biden won the state's twenty electoral votes by a margin of just 1.17%. This book is going to press just before the 2024 election, but there's little reason to think the result will be dramatically different when Trump goes up against Kamala Harris (though slow population growth means the state delivers only nineteen Electoral Votes this time).

The city of Pittsburgh, along with southeastern Allegheny County and a portion of Westmoreland County, is in Pennsylvania's 12th congressional district, currently represented by Democrat Summer Lee. Lee, a community organizer and environmental justice activist from the Mon Valley town of Braddock, is the first Black woman to represent Western Pennsylvania in the US Congress. She is known for being one of the most left-wing members of that body.

While PA-12 is considered a "safe" Democratic seat in the general election, Lee drew a tough 2024 primary challenge from entrepreneur Bhavini Patel. Patel argued that Lee's radicalism hampered her ability to work with the moderate President Biden. The biggest issue of the campaign was Patel's support for Israel versus Lee's outspoken criticism, and related questions about the candidates' abilities to handle problems of antisemitism and Islamophobia. Patel made a strong showing both in the generally conservative rural east of the district and in the Squirrel Hill neighborhood where "Stand with Israel" yard signs have outnumbered "Free Palestine" ones in the months since the October 7, 2023 surprise attack by Hamas against Israeli citizens and Israel's devastating military retaliation in Gaza. Lee did particularly well in Oakland, where students organized a Gaza Solidarity Encampment to demand that Pitt divest from companies connected to Israel, and went on to claim 60% of the primary vote.

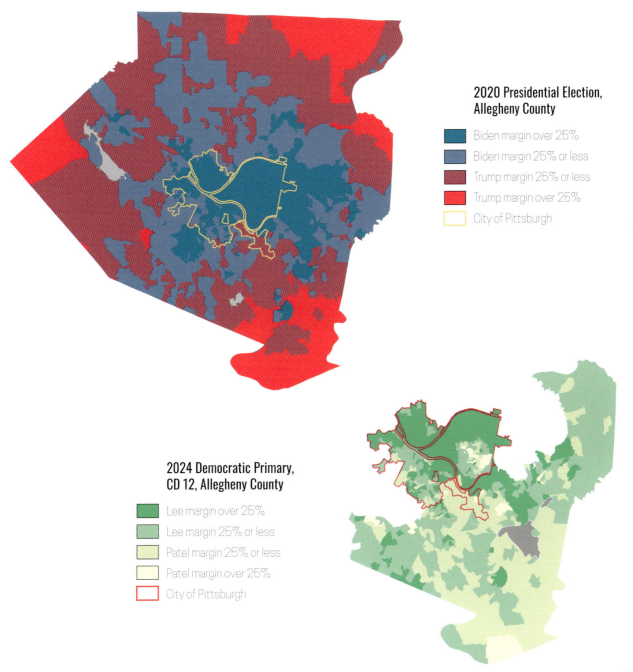

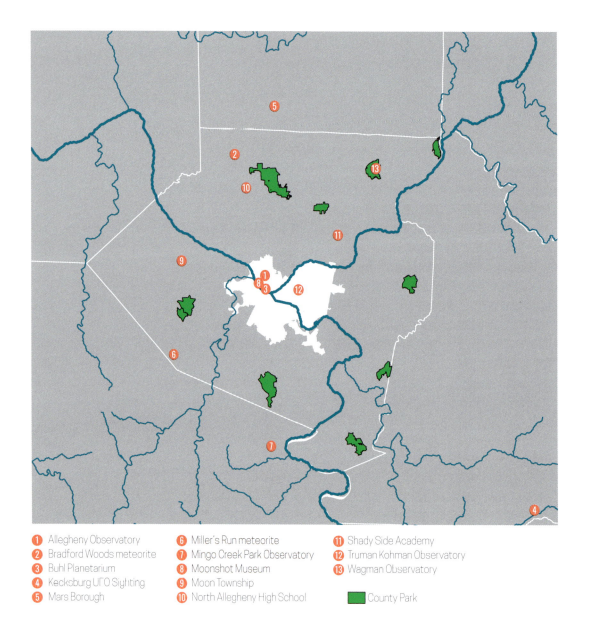

To Outer Space

Oral tradition from the Seneca recounts how human inhabitation of the world began when Sky Woman fell down from the heavens. But it wasn't until 1991 that any Yinzer actually traveled up into the heavens. That year, Shady Side Academy graduate Jerome "Jay" Apt flew the first of an eventual four space shuttle missions before returning to Pittsburgh to join the faculty of CMU. In March of 2024, Pittsburgh native and North Allegheny High School graduate Warren "Woody" Hoburg led a SpaceX crew to the International Space Station. He snapped a photo of himself waving a Terrible Towel to support the Steelers from 240 miles above the Earth.

Those of us who are not astronauts can pay a visit to the Allegheny Observatory in Riverview Park; the Wagman Observatory in Deer Lakes Park; the Mingo Creek Park Observatory; or the Truman Kohman Observatory at CMU. The Buhl Planetarium at the Kamin Science Center is also a great place to learn about the cosmos. In 2024, the Moonshot Museum opened on the Northside in a building shared with Astrobiotic, a robotics company.

Just outside Pittsburgh, you can visit Moon and Mars. Moon Township, to the west, gets its name from the crescent-shaped south bank of the Ohio River. The origin of the name Mars, to the north, may be a nod to founder Samuel Parkes's wife's interest in astronomy. Both communities have embraced their celestial associations, with an astronaut statue in Moon and a huge flying saucer in Mars. An hour's drive to the southeast, the town of Kecksburg has an annual UFO festival to celebrate a famous 1965 sighting of what some residents believe was an alien craft.

Allegheny County has a low overall number of sightings reported to the National UFO Reporting Center, at just 37.6 per 100,000 people during the 21st century. But the area has had landings from outer space. A thirty-pound iron meteorite was unearthed by a farmer along Miller's Run, south of the city, in 1850, and in 1886 a two-pound stony meteorite was found by another farmer, in Bradford Woods, north of the city, after he witnessed it fall. In 2022, a thousand-pound meteor exploded over the city, causing a shock wave that rattled residents, though no pieces reached the ground.

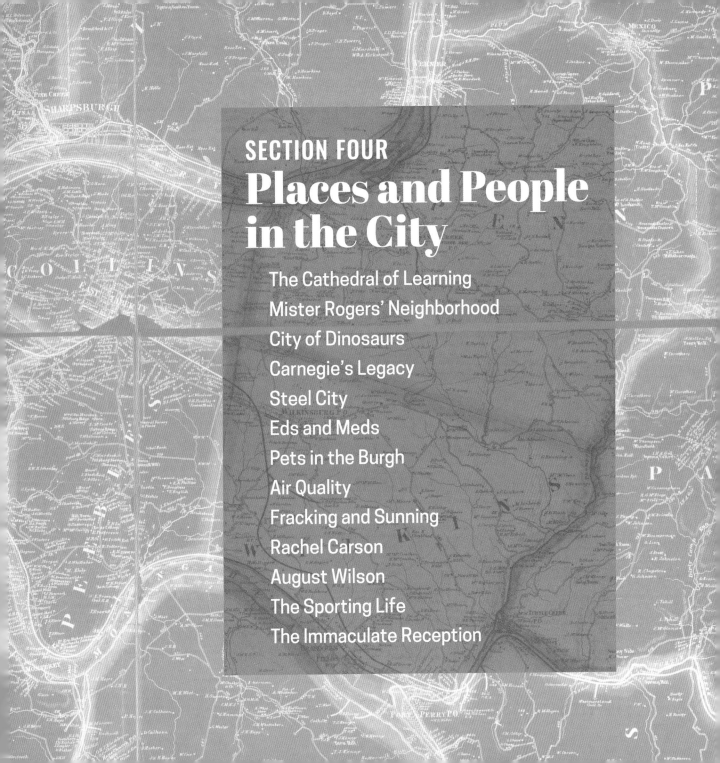

SECTION FOUR
Places and People in the City

The Cathedral of Learning
Mister Rogers' Neighborhood
City of Dinosaurs
Carnegie's Legacy
Steel City
Eds and Meds
Pets in the Burgh
Air Quality
Fracking and Sunning
Rachel Carson
August Wilson
The Sporting Life
The Immaculate Reception

①	German, 1938	⑨	Chinese, 1939	⑰	Italian, 1949	㉕	Japanese, 1999
②	Russian, 1938	⑩	Polish, 1940	⑱	English, 1952	㉖	Indian, 2000
③	Scottish, 1938	⑪	Lithuanian, 1940	⑲	Irish, 1957	㉗	Welsh, 2008
④	Swedish, 1938	⑫	Syria-Lebanon, 1941	⑳	Israel Heritage, 1987	㉘	Turkish, 2012
⑤	Early American, 1938	⑬	Greek, 1941	㉑	Armenian, 1988	㉙	Swiss, 2012
⑥	Czechoslovak, 1939	⑭	French, 1943	㉒	African Heritage, 1989	㉚	Korean, 2015
⑦	Yugoslav, 1939	⑮	Romanian, 1943	㉓	Ukrainian, 1990	㉛	Philippine, 2019
⑧	Hungarian, 1939	⑯	Norwegian, 1948	㉔	Austrian, 1996		

The Cathedral of Learning

The campus of the University of Pittsburgh is architecturally dominated by the Cathedral of Learning, a 42-story, 535-foot-tall Gothic Revival tower which is the second-tallest university building in the world. (The tallest is the main building of the University of Moscow.) The building was begun in 1926, the first classes were held in it in 1931, and it was officially dedicated in 1937. Funding for construction came from a variety of sources, including a campaign in which ninety-seven thousand schoolchildren donated a dime each. The building, called "Cathy" by some Pitt students, houses classrooms, administrative offices, study spaces, and a food court. Offices in the Cathedral include the University Chancellor and Vice Chancellor, the Provost, Frederick Honors College, and the departments of English, Philosophy, History and Philosophy of Science, Communication, Classics, Asian Studies, Linguistics, Religious Studies, Theatre Arts, and Social Work. A webcam atop the building allows the public to keep an eye on a nesting pair of peregrine falcons.

A major attraction at the Cathedral of Learning is the thirty-one "Nationality Rooms," classrooms decorated to honor the various ethnic groups that make up Pittsburgh's population. The rooms depict the traditional cultures of these groups prior to the university's founding in 1787. The first four rooms to be opened were the German, Russian, Scottish, and Swedish rooms in 1938. The newest is the Philippine room, which was dedicated in 2019. Each room's décor is based on authentic architecture and furnishings from the source culture—for example, an Asante temple courtyard for the African Heritage room, the fourteenth century Myeong-nyundang (Hall of Enlightenment) from Korea's royal academy, or a Swedish peasant cottage. In addition to serving as regular classrooms, the rooms also host lectures and cultural events related to the nationalities they showcase. Visitors can sign up for a guided tour of the nationality rooms as long as classes are not occurring.

Mister Rogers' Neighborhood

Fred McFeely Rogers (1928–2003) was a beloved children's television host who mixed compassion and imagination in his long-running PBS program *Mister Rogers' Neighborhood*. Born and raised in Latrobe, forty miles east of Pittsburgh, he made the city his home for most of his life.

Rogers was a shy child who found solace in making puppets. He attended high school in Latrobe, then earned a degree in music from Rollins College in Florida. He came to Pittsburgh to attend seminary, earning a bachelor of divinity and being ordained a minister in the Presbyterian Church. He also studied at the Pitt Graduate School of Child Development. He married his Rollins classmate, Sara Joanne Byrd, with whom he had two sons. Later in life, he came out as bisexual.

After a short stint in New York City, Rogers took a job with public TV station WQED Pittsburgh in 1953. Between 1963 and 1967, he moved to Toronto to star in his own program, *Misterogers*. He returned to WQED and began broadcasting *Mister Rogers' Neighborhood* in 1968. The show would run for more than three decades, and it continues to air in reruns.

The show was lauded for its calm pace, its educational quality, and Rogers' ability to address children's social and emotional needs. Nevertheless, Rogers didn't shy away from tough topics, producing an episode about the assassination of President Kennedy, and spending time in a kiddie pool with Officer Clemons, played by a Black actor, at the height of the civil rights movement. In 1969, he gave a powerful testimony to Congress in defense of public television.

Rogers died of stomach cancer in 2003. His family held a private funeral at Unity Cemetery in Latrobe, where he was buried in the family mausoleum. A public memorial, attended by 2,700 people, was held at Heinz Hall. Many children's entertainers, including Marc Brown (*Arthur*) and Angela Santomero (*Blue's Clues*) cite Rogers as a major influence, and his advice to "look for the helpers" regularly goes viral on social media after tragic events. He has been honored with a large bronze statue on Pittsburgh's North Shore, a "Fredosaurus Rex" dinosaur statue outside the Fred Rogers Productions office, and the Fred Rogers Center for Early Learning and Children's Media at St. Vincent College.

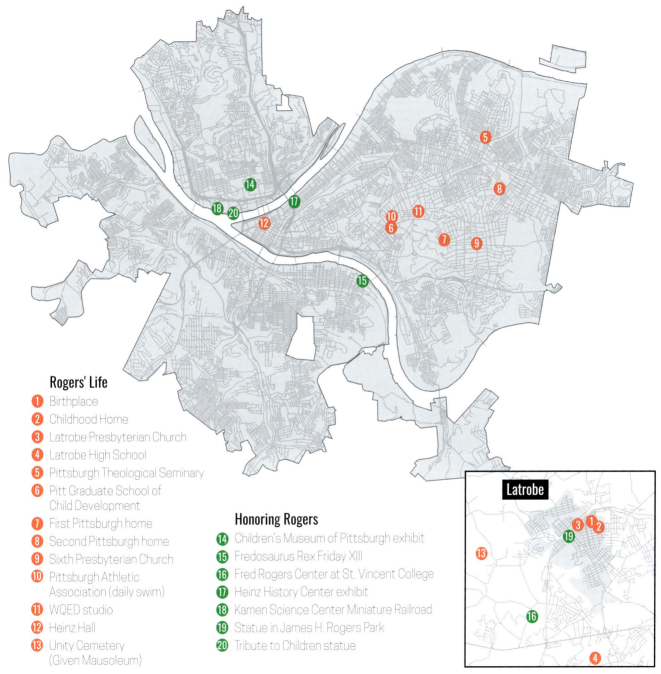

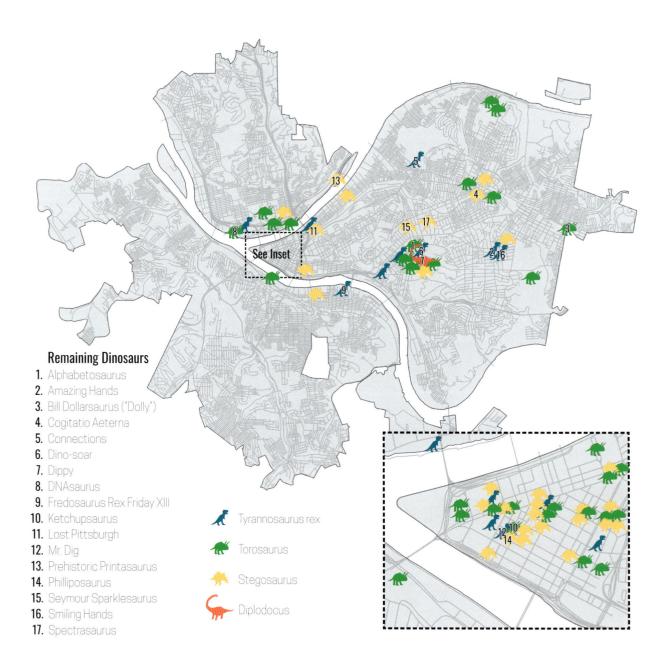

City of Dinosaurs

Pittsburgh has a long association with dinosaurs. Over the 186 million years of the Mesozoic Era, the area where the city is now located was home to a variety of dinos, including the duck-billed Hadrosaurus and the carnivorous Dryptosaurus, and possibly even the iconic Diplodocus. Unfortunately, dinosaur fossils are rare in the area because our main rock formations were laid down during the Pennsylvanian period, well before the time when dinosaurs made their home here.

Steel baron and philanthropist Andrew Carnegie took a particular interest in dinosaurs. After reading a newspaper article about the discovery of Diplodocus in Wyoming in 1898, Carnegie sent a $10,000 check to the Carnegie Museum of Natural History (CMNH), along with a note: "Dear Chancellor, buy this for Pittsburgh." The skeleton brought to the museum was given the scientific name *Diplodocus carnegii*. Since then, the museum has acquired a world-class collection of dinosaur fossils to educate the public. A life-size Diplodocus statue, affectionately nicknamed Dippy, stands outside the museum and is often given a decorative scarf for special occasions (e.g., a rainbow scarf during Pride Month).

In 2003, an art exhibit called Dinomite Days installed one hundred fiberglass statues of Tyrannosaurus, Torosaurus, and Stegosaurus all around the region. In exchange for a donation to support the CMNH, sponsors were able to decorate the statues in a variety of ways; the results included a Ketchupsaurus, in the form of a Heinz ketchup bottle, and an Alphabetasaurus, covered in letters drawn by public elementary school students. Over the subsequent two decades, many of these dinosaurs have been damaged, lost, or sold, but you can still see a few of them stalking the city. The map shows all of the dinosaurs originally placed within the city limits along with the names of the ones I've been able to confirm are still visible to the public—but dedicated dino-hunters may be able to locate a few more in hiding.

Carnegie's Legacy

Andrew Carnegie (1835–1919) came to Pittsburgh from Scotland as a boy and went on to become a titan of the steel industry and one of the richest men in the world. Depending on who you ask, his life is either a rags-to-riches tale of the American dream or a dystopian story of exploitation and the rise of one of the Gilded Age's greediest robber barons. Either way, Carnegie made a lot of money, and he gave much of it away—about 90 percent of his fortune—through philanthropic efforts that have left his name scattered across the Pittsburgh landscape.

The most famous of Carnegie's projects are the libraries. Some three thousand Carnegie Libraries were established around the world, including eight in Pittsburgh, where the modern Carnegie Library of Pittsburgh system includes twenty locations. He also founded the Carnegie Institute of Technology, which later merged with the Mellon Institute of Industrial Research to become Carnegie Mellon University. The Carnegie Museums were also beneficiaries of his philanthropy, although in 2024, Carnegie's name was replaced by that of Daniel G. and Carole L. Kamin in the title of the Science Center after the Kamins made their own major donation. The borough of Carnegie, which borders Pittsburgh to the west, and the adjoining city neighborhood of East Carnegie, are also named after him.

Carnegie's philanthropy was largely successful in securing a positive legacy, associating his name in the public mind with civic and charitable institutions that enrich public life, and establishing a model for other ultra-rich donors such as Bill Gates. Critics have, however, pushed back against the elitism inherent to the idea that one rich man, even a well-meaning one, should be in control of how so much money is spent and which charitable causes deserve funding. Carnegie wrote that working-class people couldn't be trusted with higher wages because they would waste them on "indulgence of appetite" in contrast to his own enlightened altruism. Today, nonprofits and community groups often have to pander to the interests of donors, sometimes at the expense of their mission. Moreover, the benefits of philanthropy must be weighed against harms done in the process of accumulating wealth, such as Carnegie's role in violently crushing the 1892 Homestead Strike, which left seven steelworkers and three Pinkerton officers dead.

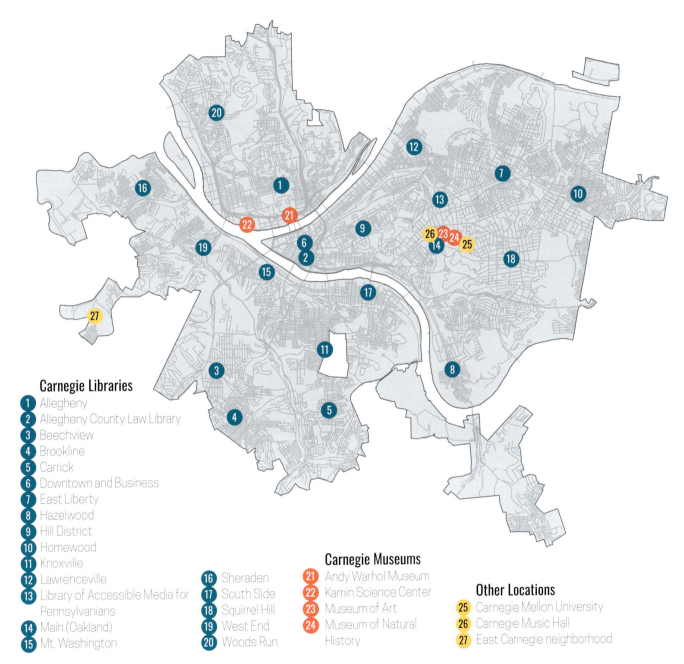

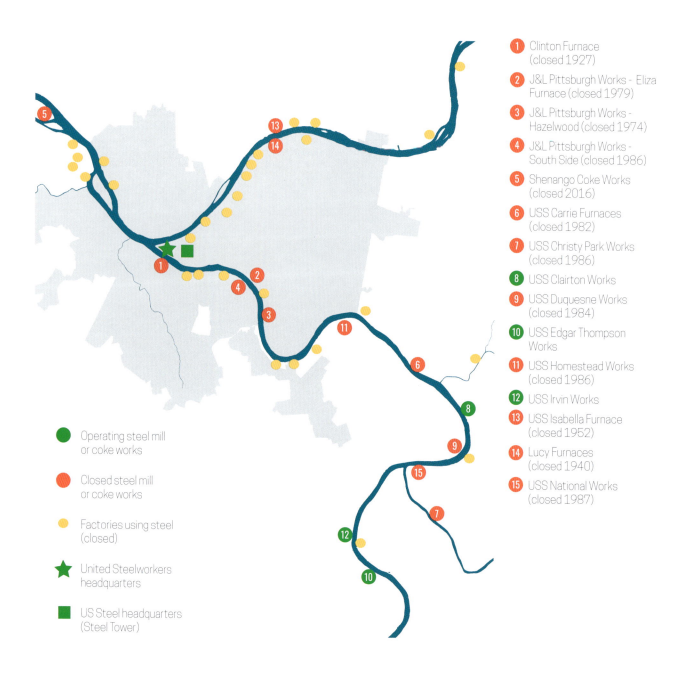

Steel City

Pittsburgh's three rivers enabled easy transportation of iron ore and coal from nearby mines, and in the early 1800s, ironworks began to spring up along the riverbanks. In the 1870s, Andrew Carnegie brought the Bessemer process—a new technology that made large-scale steel production more feasible—to his mill in Lawrenceville, and steel soon became a core component of the Pittsburgh economy. Other factories sprang up in and around the city for access to steel to manufacture goods ranging from railroad cars to springs.

The 1901 merger of Carnegie Steel, Federal Steel, and National Steel formed the powerhouse US Steel corporation (USS), which would dominate the steel industry, and life in Pittsburgh, for nearly a century. At its peak, the Pittsburgh area was producing 60 percent of the nation's steel. In 1971, the company opened its headquarters in the Steel Tower, Pittsburgh's tallest building. Heavy competition from foreign companies would eventually knock USS off its throne, however, and a wave of plant closures swept the Pittsburgh area in the 1980s. A similar fate also befell Jones & Laughlin, Pittsburgh's number two steel company.

The city's mills also fostered the rise of organized labor, as workers came together to fight for better wages and working conditions, sometimes facing violent repression by company owners. Unionization in Pittsburgh's iron and steel industry dates back to the 1858 founding of the Sons of Vulcan. Eventually unions throughout the country united to form the United Steelworkers (USW) in 1942. The union expanded its organizing efforts into other parts of the workforce, including workers at the Carnegie Museums, Persad Center, and several universities.

Today, three steel mills remain active in the Monongahela Valley south of the city, and USS and USW maintain their headquarters in Downtown Pittsburgh. The Clairton Coke Works has faced years of community activism and litigation over its impact on local air quality, culminating in a 2024 settlement with the EPA in which it agreed to permanently shutter one of its batteries. In 2024, a deal was announced for the Japanese company Nippon Steel to acquire USS. National and local politicians and the USW have vowed to block the deal.

Eds and Meds

The decline of the steel industry in the latter half of the twentieth century left Pittsburgh searching for a new economic base. The city's largest employers today are educational and medical institutions. Some see this focus on "eds and meds" as a way for the city to chart a viable path to a post-industrial future.

Pittsburgh is home to eleven institutions of higher education. Some, like the University of Pittsburgh and Carnegie Mellon University, are research powerhouses, generating groundbreaking scientific discoveries and training researchers who go on to work in the city's growing technology sector, for companies like Google and Uber. Other colleges and universities play a crucial role in educating large numbers of students for a variety of careers in medicine, business, and the arts.

Health care in Pittsburgh is dominated by two major players: the University of Pittsburgh Medical Center (UPMC) and Highmark, which owns the Allegheny Health Network (AHN). The downtown skyline features the US Steel Tower, which today bears a huge UPMC logo, and the multicolored lights on top of the Highmark Building. Conflicts between the two health giants and their associated insurance programs can complicate patients' search for appropriate care.

A challenge for the city is that property owned by nonprofit medical and educational organizations is exempt from property taxes, depriving the city of revenue it would have received had that land been in private ownership even while these institutions benefit from city services like road maintenance and fire protection. Tax-exempt properties make up more than a third of parcels in the city, and Mayor Ed Gainey's administration has filed lawsuits to challenge over a hundred of them. Leaders of these institutions contend that tax exemptions enable them to afford to provide important services to the community. An alternative arrangement, in which major nonprofits provide "payment in lieu of taxes" to local government, is used in other cities but not currently in Pittsburgh.

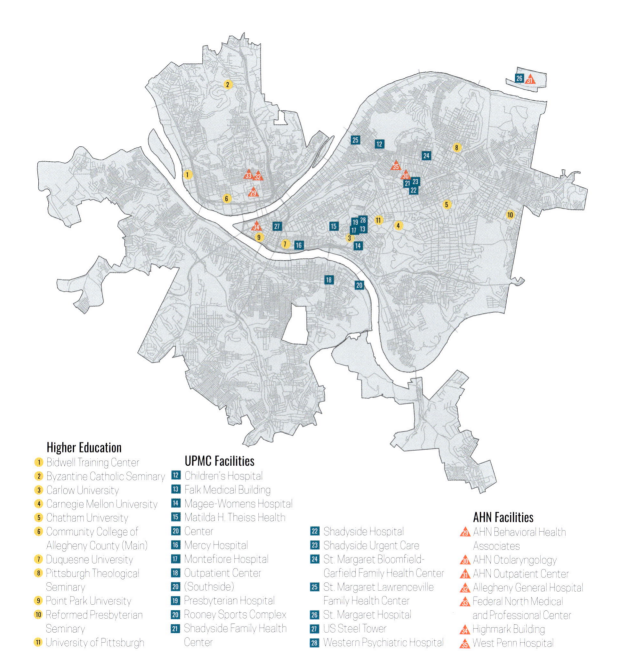

Higher Education
1. Bidwell Training Center
2. Byzantine Catholic Seminary
3. Carlow University
4. Carnegie Mellon University
5. Chatham University
6. Community College of Allegheny County (Main)
7. Duquesne University
8. Pittsburgh Theological Seminary
9. Point Park University
10. Reformed Presbyterian Seminary
11. University of Pittsburgh

UPMC Facilities
12. Children's Hospital
13. Falk Medical Building
14. Magee-Womens Hospital
15. Matilda H. Theiss Health Center
16. Mercy Hospital
17. Montefiore Hospital
18. Outpatient Center (Southside)
19. Presbyterian Hospital
20. Rooney Sports Complex
21. Shadyside Family Health Center
22. Shadyside Hospital
23. Shadyside Urgent Care
24. St. Margaret Bloomfield-Garfield Family Health Center
25. St. Margaret Lawrenceville Family Health Center
26. St. Margaret Hospital
27. US Steel Tower
28. Western Psychiatric Hospital

AHN Facilities
29. AHN Behavioral Health Associates
30. AHN Otolaryngology
31. AHN Outpatient Center
32. Allegheny General Hospital
33. Federal North Medical and Professional Center
34. Highmark Building
35. West Penn Hospital

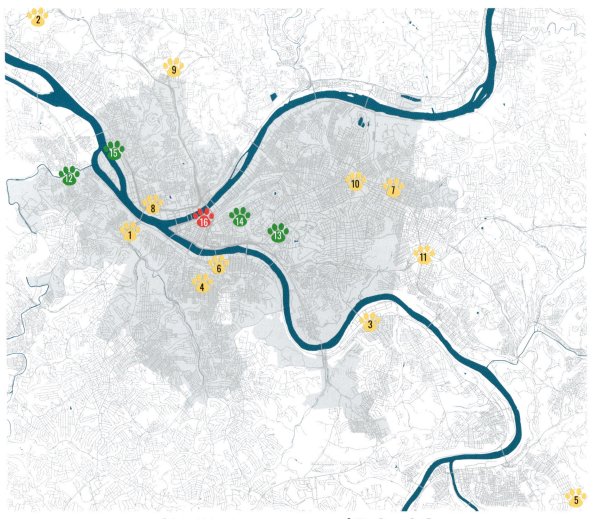

 Shelter or other facility
 TNR site
 Other location

1. Animal Advocates
2. Animal Friends
3. Animal Lifeline Thrift Store
4. Biggies Bullies
5. Gray Paws Sanctuary
6. Heart of Glass Rescue
7. Humane Animal Rescue East
8. Humane Animal Rescue North
9. Kitty Queen Cat Rescue
10. Petland Village of Eastside
11. Rescue and Relax
12. Cat Ears Revolution
13. Oakland TNR Coalition
14. Pittsburgh Hill District Cats
15. Pittsburgh Prison Cats
16. Anthrocon, David L. Lawrence Convention Center

Pets in the Burgh

My name may be on the deed, but the real owners of my house are my two cats, Monks (black and white) and Rodrigo (brown tabby). Both of them were "foster fails" we took in as foster kittens from Humane Animal Rescue but couldn't bear to give back. (We've also fostered dozens of kittens that went on to forever homes all around the region.)

The Pittsburgh area is home to more than a dozen animal shelters and rescue operations, some focused on specific species or breeds, and others caring for all kinds of pets. There are also several organizations dedicated to "trap-neuter-release" (TNR) programs aimed at managing feral cat colonies. Among the most interesting of these colonies are the Pittsburgh Prison Cats, who live in the former Western State Penitentiary along the Ohio River. Not shown on the map are several organizations—including Furkid Rescue, Paws Across Pittsburgh, PEARL Parrot Rescue, Rabbit Wranglers, and Trash Cat Rescue—that don't have a fixed location but rather maintain a network of volunteer foster homes.

Every Fourth of July weekend, the David L. Lawrence Convention Center hosts Anthrocon, one of the world's largest gatherings of "furries"—people who like to dress up as their animal alter ego, or "fursona"—with over thirteen thousand attendees in 2023. The conference raises money for a local animal charity; in 2023, it raised $52,000 for Rabbit Wranglers.

Air Quality

Ever since an 1866 *Atlantic Monthly* article described Pittsburgh as "hell with the lid taken off," the city has been known for its poor air quality. The decline of heavy industry and increases in environmental regulation over the last half century have dramatically cleaned up the city's air, but Pittsburgh still ranks near the bottom for US cities in terms of air quality.

The EPA sets national standards for six major "criteria air pollutants." In 2022, the county's air monitors recorded two days of exceeding federal standards for ozone (O3) pollution and eight days exceeding the standard for small particulate matter (PM2.5). These are substantial improvements from the past—for example, in 2002, the county had twenty-two days exceeding the ozone standard. A report from the American Lung Association ranked Pittsburgh the nineteenth worst metro area for PM2.5, better than Detroit but worse than Cleveland or Buffalo. Air monitoring stations in the Mon Valley tend to record the worst air pollution in the county, with the Clairton Coke Works consistently the biggest single source.

The Toxics Release Inventory is a federal program that monitors the release of other hazardous pollution from facilities around the country. Most TRI sites in Pittsburgh lie along the rivers, in the city's traditional industrial corridors. In 2022, the biggest emitting facilities in the area were the McConway and Torley rail car coupling factory in Lawrenceville, and the Calgon Carbon Corp. and Ineos Composites chemical plants on Neville Island.

A new form of air pollution began to hit Pittsburgh in the summer of 2023—wildfire smoke. Massive fires burning in western Canada produced smoke that was brought to Pennsylvania by westerly winds, turning the sky orange and impacting residents' health. Scientists warn that climate change will make such large fires, and their smoke, a more common occurrence.

Air pollution can cause, or aggravate, a variety of health conditions, with asthma being one of the most immediately noticeable. Allegheny County's asthma rate is 10 percent, as compared to 8 percent nationally. Studies have shown that asthma rates are higher for people who are Black, low-income, or live near polluting facilities.

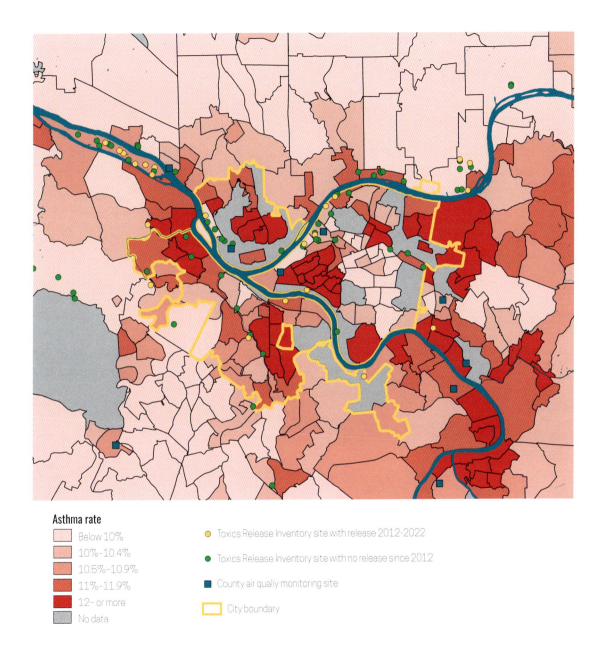

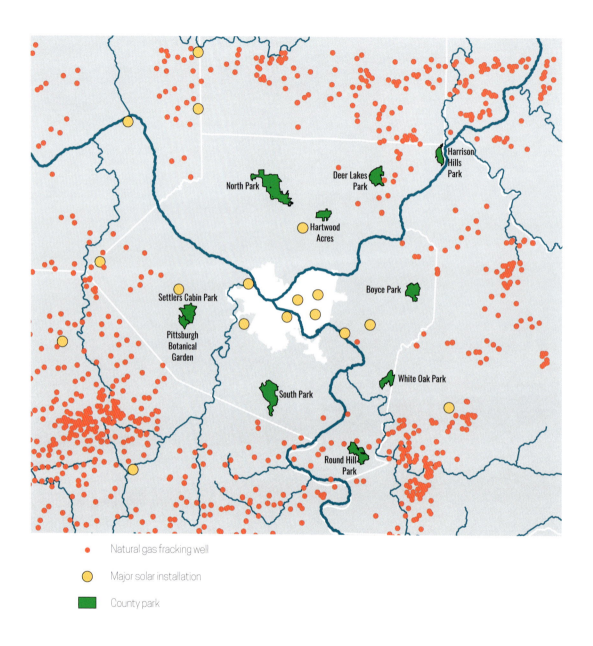

Fracking and Sunning

Right below Pittsburgh lies the Marcellus Shale, a vast rock formation containing natural gas that was, until recently, too difficult to extract. But new methods of hydraulic fracturing—commonly called "fracking"—have allowed companies to extract this material economically.

Fracking has been highly controversial in Pennsylvania for years. Boosters point to its potential to create jobs to replace those lost due to the decline of the region's coal and steel industries. They also tout the environmental benefits of replacing coal and oil with cleaner-burning natural gas. Critics, however, point to the severe water pollution, increased traffic and road-building, and even earthquakes that can result from the practice. Moreover, while burning natural gas produces fewer carbon emissions than other fossil fuels, it still contributes more to climate change than non-fossil fuel energy sources.

Generally, fracking has been more common in Pittsburgh's surrounding counties and the edges of Allegheny County than in the city itself or its immediate suburbs. In 2022, the county council overrode the executive's veto to pass a ban on fracking in the county's ten parks.

Pennsylvania lags behind other states in building solar and wind energy. Only 3 percent of the state's energy currently comes from renewable sources. Unlike fracking, solar installations have been more popular in the city and its immediate surroundings than in more rural areas. Currently, the Pennsylvania Solar Center—a nonprofit that works to promote renewable energy—identifies eleven large solar installations in Allegheny County.

Rachel Carson

Rachel Carson (1907–1964) is widely credited with helping to spark the modern environmental movement with her writing about nature—especially her 1962 classic, *Silent Spring*, which warned of the dangers of overuse of pesticides. She endured a barrage of sexism from critics who thought a woman couldn't speak on issues of science, and homophobia from those who noticed that she never married and had a close relationship with a woman, Dorothy Freeman.

Carson grew up in Springdale, just upstream of Pittsburgh on the Allegheny. She went to high school across the river in Parnassus (now part of New Kensington), then enrolled at the Pennsylvania College for Women (now Chatham University), where she studied biology and wrote for the student newspaper. After a summer at the Marine Biological Laboratory at Woods Hole, she enrolled at Johns Hopkins University to get her master's degree in marine biology, finishing in 1932. She then went to work for the US Bureau of Fisheries, writing public education materials. She also wrote extensively for newspapers and magazines, and she published her first book, *Under the Sea Wind*, in 1941. In 1953, she moved to the coast of Maine to be a full-time writer. Four years later, she moved back to Maryland to be closer to family; there, she composed *Silent Spring*. The book, which forced the nation to confront the harms caused by overuse of pesticides such as DDT, made Carson famous but also sparked a fierce backlash from the chemical industry. The title comes from the vivid opening vignette of a world gone silent because pesticides have killed all of the songbirds, an image that helped readers connect with what might otherwise have been dry environmental science statistics.

Carson died in 1964 after a battle with breast cancer. Half of her ashes were buried next to her mother in Rockville, Maryland, while the other half were scattered in the bay near her old home in Maine. Her personal papers are archived at the Yale Library, and a variety of locations around the country are named in honor of her. In Pittsburgh, one of the "Three Sisters" bridges between the North Shore and Downtown bears her name; the Rachel Carson Trail winds across the hilly northern part of the county. Her homes in Springdale and Colesville are both designated historic landmarks.

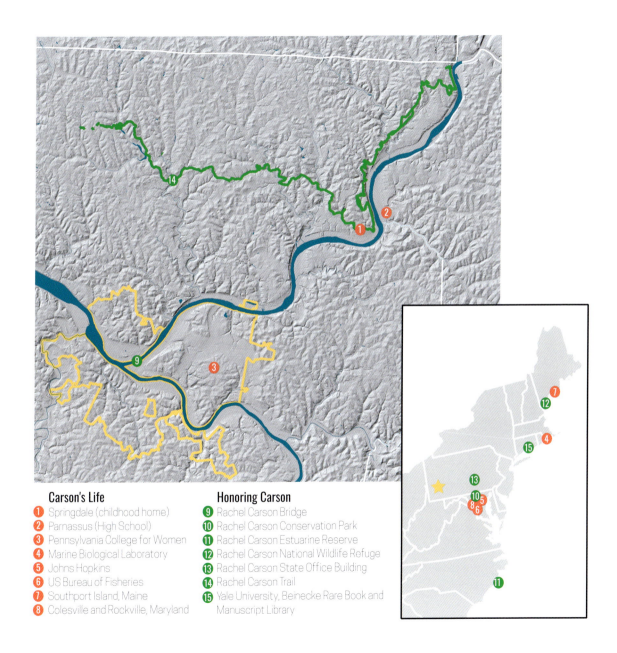

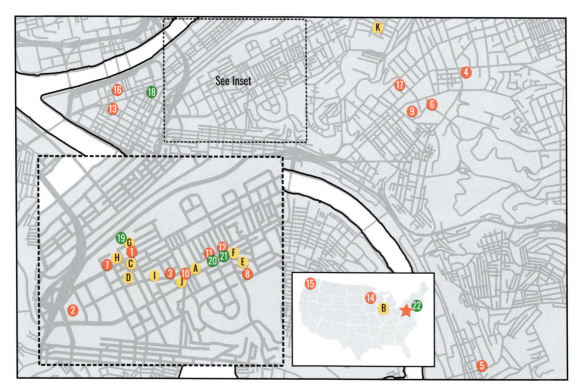

Wilson's Life
1. Childhood Home (August Wilson House)
2. Holy Trinity School
3. Carnegie Library (Wylie)
4. Central Catholic
5. Gladstone School
6. Carnegie Library (Main)
7. 85 Crawford (Adult Home)
8. Black Horizon Theater (A. Leo Weil Elementary School)
9. Kuntu Writers Workshop (Pitt Dept. of Africana Studies)
10. New Granada Theater
11. Crawford Grill No. 2
12. Eddie's Restaurant
13. Original Oyster House
14. St. Paul (1978-1990)
15. Seattle (1990-2005)
16. Pittsburgh Public Theater
17. Soldiers and Sailors Memorial Hall

Honoring Wilson
18. August Wilson African American Cultural Center
19. August Wilson Park
20. August Wilson Room at Carnegie Library
21. Mural at Black Beauty Lounge
22. August Wilson Theatre

Pittsburgh Cycle
A. *Jitney* (1982): Westbrook Jitney Station
B. *Ma Rainey's Black Bottom* (1984): Chicago
C. *Fences* (1985): Troy and Rose Maxson's backyard
D. *Joe Turner's Come and Gone* (1986): Seth and Bertha Holly's Boarding House
E. *The Piano Lesson* (1987): Diamond's Five and Ten
F. *Two Trains Running* (1990): Memphis's Diner
G. *Seven Guitars* (1995): Vera, Louise, Floyd, and Hedley's house
H. *King Hedley II* (1999): Ruby, King, and Tonya's house
I. *Gem of the Ocean* (2003): Aunt Ester's House
J. *Radio Golf* (2005): Bedford Hills Redevelopment Office
K. *Fences* movie (2016): Filming location

August Wilson

August Wilson (1945–2005) is the city's most famous playwright. His plays address life in the city's Black communities over the course of the twentieth century. Wilson is widely recognized as a pivotal voice in the development of Black theater in the US, and two of the ten plays in the "Pittsburgh Cycle," *Fences* (1987) and *The Piano Lesson* (1990), won the Pulitzer Prize.

Wilson was born in the Hill District, then moved with his mother to Hazelwood. He bounced between several high schools before dropping out, but he continued to self-educate at the Carnegie Library, which awarded him an honorary diploma in 1999. In 1968, he cofounded the Black Horizon Theater, which staged some of his early work. Then, in 1976, he cofounded the Kuntu Writers Workshop to assist Black writers. He would sometimes work on his plays at several of his favorite restaurants in the city, such as the Crawford Grille, Eddie's, and Original Oyster House. Wilson moved to St. Paul, Minnesota, in 1978 and eventually to Seattle in 1990, but even after leaving Pittsburgh, he continued to write plays based on his experiences there. He passed away in 2005 after a short bout with liver cancer. His memorial service was held at Soldiers and Sailors Hall in Oakland, and he is buried in Greenwood Cemetery in O'Hara Township. In 2017, the Pittsburgh Public Theater staged the first performance of the entire Pittsburgh Cycle of plays.

Wilson accumulated a long list of honors in his life and after. In addition to his Pulitzer wins, he won the Tony Award for Drama twice, and the 2016 film adaptation of *Fences* was nominated for an Oscar. Today, Downtown hosts the prominent August Wilson Center for African American Culture, a park in the Hill District is named for him, and his name graces the former Virginia Theatre on Broadway, the first Black person to be so honored.

The Sporting Life

Pittsburgh is a sports town. Most of the attention goes to the "big three" major teams—the Pirates (baseball), Steelers (football), and Penguins (hockey). But several other notable professional teams have also called the city home, including the Homestead Grays of Negro League baseball; the Passion, a women's football team that plays their games at a stadium in the South Side; the Hornets, a minor-league hockey team that was active until 1967, when the Penguins came to town; and the Riverhounds soccer squad. Many shorter-lived teams have also played in the city, as well as a variety of amateur clubs and teams representing the city's educational institutions, but for this map I'm only considering professional teams that played in Pittsburgh for at least ten seasons.

Unlike some cities that send their sports teams to distant suburban stadiums, Pittsburgh has a long history of having its teams play in the heart of the city. The Steelers and Pirates shared a dual-use stadium for most of the twentieth century, first at Forbes Field in Oakland and then at Three Rivers Stadium on the North Shore. Currently, the Steelers play at Acrisure Stadium (formerly Heinz Field) and the Pirates at PNC Park.

The Penguins have always made their home in the Lower Hill area, first at the Civic Arena and now at PPG Paints Arena (formerly the Consol Energy Center). The development of these arenas was highly controversial, as they came about through an urban renewal project that wiped out what had been a vibrant Black neighborhood and cut the Hill District off from Downtown. The team's ownership is in ongoing talks with the city and local neighborhood groups about redevelopment plans that would benefit the community and undo some of the historic damage. To that end, Frankie Pace Park was opened in 2021, and an agreement was signed in 2024 to construct a music venue on the condition that the developer will hire a diverse workforce and book performances at the nearby New Grenada Theater, a historic hub of the city's Black culture.

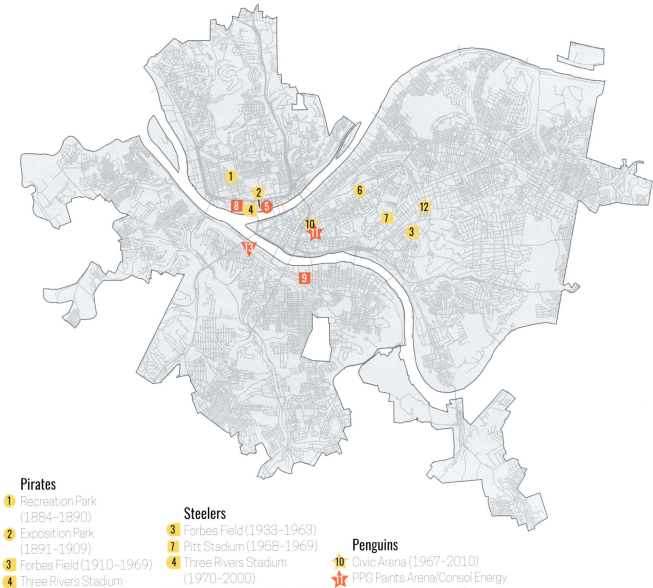

Pirates
1. Recreation Park (1884–1890)
2. Exposition Park (1891–1909)
3. Forbes Field (1910–1969)
4. Three Rivers Stadium (1970–2000)
5. PNC Park (2001–present)

Homestead Grays
3. Forbes Field (1929–1950)
6. Greenlee Field (1932–1937)

Steelers
3. Forbes Field (1933–1963)
7. Pitt Stadium (1958–1969)
4. Three Rivers Stadium (1970–2000)
8. Acrisure Stadium/Heinz Field (2001–present)

Passion
9. George K. Cupples Stadium (2006–present)

Penguins
10. Civic Arena (1967–2010)
11. PPG Paints Arena/Consol Energy Center (2010–present)

Hornets
12. Duquesne Gardens (1936–1956)
10. Civic Arena (1961–1967)

Riverhounds
13. Highmark Stadium (2013–present)

123

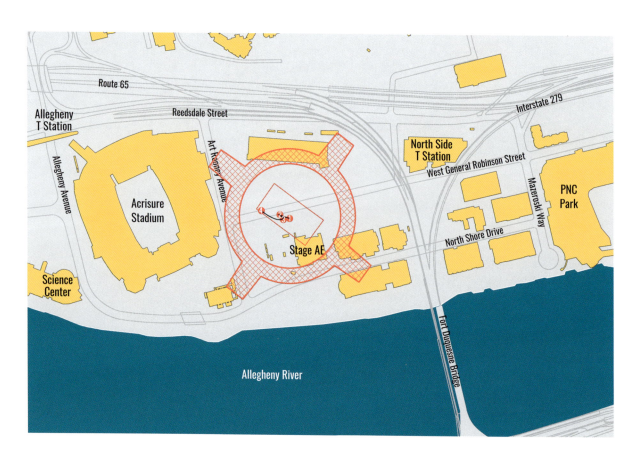

The Immaculate Reception

The greatest play in Pittsburgh Steelers history—arguably the greatest play in NFL history—happened on December 23, 1972, at Three Rivers Stadium on Pittsburgh's North Shore. With twenty-two seconds to go in the divisional playoff game against the Oakland Raiders, the Steelers were down 7–6 and facing a fourth down. Quarterback Terry Bradshaw dropped back and fired a pass to running back John Fuqua, but it bounced off Oakland safety Jack Tatum's helmet. This could have been the end of the season for the Steelers, but miraculously, running back Franco Harris lunged for the ball and caught it just before it hit the ground. Harris flew down the field forty-four yards for a game-winning touchdown. Sportscaster Myron Cope—taking a suggestion from listener Sharon Levosky's friend Michael Ord—dubbed the play the "Immaculate Reception," a pun on the Catholic dogma of the Immaculate Conception, which captured the sense of divine blessing that Steelers fans felt.

Harris's catch secured the Steelers their first-ever playoff victory. Though the Steelers would go on to lose the subsequent AFC Championship game to the Miami Dolphins, the following years saw the team's 1970s golden age, with four Super Bowl wins in six years.

The play remains controversial today, as some football fans believe the ball hit Fuqua rather than Tatum, or that it touched the ground before Harris reeled it in, either of which should have ended the play. Some detractors have even labeled it the "Immaculate Deception." But in the days before instant replay, the referees on the field allowed the play to stand, and it became NFL legend.

Three Rivers Stadium was decommissioned in 2000, but a historical marker has been erected in the parking lot of Acrisure Stadium, with a bronze footprint showing exactly where Harris stepped as he caught the ball.

Sources

Historical Maps
Fort Du Quesne, now Pittsburgh, and its Environs, *The Scots Magazine*, January 1759, https://www.mapsofpa.com/pitts/1759scotmag.jpg

Clarence M. Busch, Report of the Commission to Locate the Site of the Frontier Forts of Pennsylvania, Volume Two, 1896, http://www.usgwarchives.net/pa/1pa/1picts/frontierforts/ff26.html

J Melish, View of the Country Round Pittsburg, Travels in the United States of America, in the Years 1806 & 1807, and 1809, 1810 & 1811, 1812, https://www.mapsofpa.com/pitts/1812_1616.jpg

Augustus Hani, Woods' New Map of Pittsburgh, Alleghey and Surroundings, 1867, https://www.mapsofpa.com/pitts/1867-8530.jpg

T.M. Fowler, Pittsburgh, Pennsylvania, 1902, https://www.mapsofpa.com/pitts/locfowler.jpg

Works Progress Administration, Bituminous Coal Maps of Pennsylvania, 1934-38, https://libraries.psu.edu/about/collections/bituminous-coal-mine-maps-pennsylvania

US Geological Survey, Topographic Maps

Section One: Situating the City

City of Hills
Pittsburgh Geological Society

Allegheny County GIS Department https://openac-alcogis.opendata.arcgis.com/

Quinn Glabicki, Slip sliding away: Federal funds buy out Pittsburgh homes under threat from landslides, *Public Source*, Jan 8 2024, https://www.publicsource.org/landslides-pittsburgh-mount-washington-federal-funds-mayor-gainey-traffic/

City of Four Rivers?
Margaret J. Krauss, How Pittsburgh's Ancient Aquifer Busts The 'Fourth River' Myth, WESA, July 24 2015, https://www.wesa.fm/archives/2015-07-24/how-pittsburghs-ancient-aquifer-busts-the-fourth-river-myth

A. M. Piper, Ground water in southwestern Pennsylvania, Pennsylvania Geological Survey, 1933, https://ngmdb.usgs.gov/Prodesc/proddesc_32935.htm

Upstream Pittsburgh, https://upstreampgh.org/pittsburghs-lost-and-forgotten-streams/

Jaödeogë': A City on Indigenous Land
Jordan Engel, Turtle Island Decolonized, Decolonial Atlas, 2023, https://decolonialatlas.wordpress.com/turtle-island-decolonized/

Council of Three Rivers American Indian Center, https://www.cotraic.org/

Katie Blackley, Who lived here first? A look at Pittsburgh's Native American history, WESA, December 18 2018, https://www.wesa.fm/arts-sports-culture/2018-12-18/who-lived-here-first-a-look-at-pittsburghs-native-american-history

Pail A.W. Wallace, Indian Paths of Pennsylvania, Pennsylvania Historical and Museum Commission, 1965.

A Growing City
Pittsburgh City Archives, https://twitter.com/PghArchives/status/1227610730645590016

Diana Nelson Jones, The day the City of Allegheny disappeared, *Post-Gazette*, December 9 2007, https://www.post-gazette.com/local/city/2007/12/09/The-day-the-City-of-Allegheny-disappeared/stories/200712090229

Charlie Wolfson, With annexation 'dead,' Wilkinsburg looks toward possible home rule and tax changes, *Public Source*, August 10 2023, https://www.publicsource.org/wilkinsburg-pittsburgh-home-rule-taxes-annexation-local-government/

Shrinking and Sprawling
US Census, https://data.census.gov/

Smart Growth America, Measuring Sprawl 2014, https://smartgrowthamerica.org/wp-content/uploads/2016/08/measuring-sprawl-2014.pdf

Coal City
City of Pittsburgh, Undermined Areas, https://pghgishub-pittsburgh-pa.opendata.arcgis.com/datasets/428f48cd3ba540339ab-3d2afc94d65a9/about

First Mining of Pittsburgh Coal, Explore PA History, http://explorepahistory.com/hmarker.php?markerId=1-A-2C5

Susan J. Trewalt et al., A Digital Resource Model of the Upper Pennsylvanian Pittsburgh Coal Bed, Monongahela Group, Northern Appalachian Basin Coal Region, USGS Professional Paper 1625-C, 2000, https://pubs.usgs.gov/pp/p1625c/CHAPTER_C/CHAPTER_C.pdf

Sarah Schneider, Pittsburgh's long history with coal mining isn't so buried in the past, WESA, March 23 2017, https://www.wesa.fm/environment-energy/2017-03-23/pittsburghs-long-history-with-coal-mining-isnt-so-buried-in-the-past

Sister Cities

Sister Cities Association of Pittsburgh, https://www.sistercitiespgh.org/

Cambell Robertson, Thousands of Miles From Wuhan, a U.S. City Is Shaken by Coronavirus, The New York Times, February 6, https://www.nytimes.com/2020/02/06/us/pittsburgh-wuhan-coronavirus.html

Harrison Cann, Pittsburgh and Glasgow approach one-year sister-city anniversary, *City and State PA*, October 6 2021, https://www.cityandstatepa.com/politics/2021/10/pittsburgh-and-glasgow-approach-one-year-sister-city-anniversary/364391/

Katie Blackley, Donetsk, a city in a separatist region of Ukraine, is one of Pittsburgh's 'sister cities', WESA, February 24 2022, https://www.wesa.fm/arts-sports-culture/2022-02-24/donetsk-a-city-in-a-separatist-region-of-ukraine-is-one-of-pittsburghs-sister-cities

Postal Pittsburgh

United States ZIP Codes, https://www.unitedstateszipcodes.org

Pittsburgh with an *h*

Wikipedia, Pittsburgh, https://en.wikipedia.org/wiki/Pittsburg

Connor Sites-Bowen, Introducing H Day, a New Holliday for Pittsburgh, *The Glassblock*, July 13 2016, http://theglassblock.com/2016/07/13/introducing-h-day/

Famous Firsts

Visit Pittsburgh, Pittsburgh Facts & Trivia, https://www.visitpittsburgh.com/media/press-kit/pittsburgh-facts-trivia/

Pitt Department of Medicine, Pittsburgh Fun Facts & Firsts, https://hospitalist.pitt.edu/pittsburgh-fun-facts-firsts/

Positively Pittsburgh, Pittsburgh Origins, https://positivelypittsburgh.com/pittsburgh-firsts/

Yinz N'At

City Paper, Pittsburghese Dictionary: How to Talk Like a Yinzer, *Pittsburgh City Paper*, June 9 2021, https://www.pghcitypaper.com/specials-guides/pittsburghese-dictionary-how-to-talk-like-a-yinzer-19623370

Pittsburghese.com http://www.pittsburghese.com/

Wikipedia, Western Pennsylvania English, https://en.wikipedia.org/wiki/Western_Pennsylvania_English

Heavy.ai Tweetmap https://www.heavy.ai/demos/tweetmap

Section Two: Getting around the City

Highways, Tunnels, and Belts

Wikipedia, Allegheny County belt system, https://en.wikipedia.org/wiki/Allegheny_County_belt_system

Allegheny, Ohio, and Monongahela Watersheds

USGS, Watershed Boundary Dataset, https://www.usgs.gov/national-hydrography/watershed-boundary-dataset

Wikipedia, List of longest rivers of the United States (by main stem), https://en.wikipedia.org/wiki/List_of_longest_rivers_of_the_United_States_(by_main_stem)

Hannah Wyman, Experts: Pittsburgh water safe to drink following East Palestine train derailment, *Pittsburgh Union Progress*, February 19 2023, https://www.unionprogress.com/2023/02/19/experts-pittsburgh-water-safe-to-drink-following-east-palestine-train-derailment/

Islands of Pittsburgh

Janice Lane Palko, The River Islands of Pittsburgh, *Positively Pittsburgh*, 2022, https://positivelypittsburgh.com/the-ris-islands-of-pittsburgh/

Sue Morris, Pittsburgh's Lost Islands, *The Historical Dilettante*, May 11 2016, http://historicaldilettante.blogspot.com/2016/05/things-that-arent-there-any-more.html

City of Bridges

Mark Houser, Does Pittsburgh Really Have More Bridges Than Any Other City?, Pittsburgh Magazine, March 16 2022, https://www.pittsburghmagazine.com/does-pittsburgh-really-have-more-bridges-than-any-other-city/

City of Pittsburgh, Pittsburgh Bridges, https://data.wprdc.org/dataset/pittsburgh-bridges

Ed Blazina, State makes progress in fixing bridges, but report warns that may change without additional funding, *Pittsburgh Union Progress*, June 20 2024, https://www.unionprogress.com/2024/06/20/state-makes-progress-in-fixing-bridges-but-report-warns-that-may-change-without-additional-funding/

The Pittsburgh Left

Chris Potter, I'm still wondering about the Pittsburgh Left. No one can tell me where it comes from, *Pittsburgh City Paper*, June 22 2006, https://www.pghcitypaper.com/columns/im-still-wondering-about-the-pittsburgh-left-no-one-can-tell-me-where-it-comes-from-1334231

Katie Blackley, How the Pittsburgh Left became embedded in city driving, WESA, July 28 2022, https://www.wesa.fm/development-transportation/2020-07-28/how-the-pittsburgh-left-became-embedded-in-city-driving

Underground Railroad

Julia Maruca, Tracing the Underground Railroad in Pittsburgh through these 9 historical sites, *Pittsburgh City Paper*, July 14 2020, https://www.pghcitypaper.com/news/tracing-the-underground-railroad-in-pittsburgh-through-these-9-historical-sites-17645008

Heinz History Center, From Slavery to Freedom, https://www.heinzhistorycenter.org/whats-on/history-center/exhibits/from-slavery-to-freedom/

Kiley Koscinski, Pittsburgh City Council wants to support two Juneteenth festivals amid criticism from organizer, WESA, May 29 2024, https://www.wesa.fm/politics-government/2024-05-29/pittsburgh-city-council-wants-to-support-two-juneteenth-festivals-amid-criticism-from-organizer

Rob Taylor Jr., Some 30 years ago, WAMO began 'Juneteenth' celebrations in Pittsburgh, *New Pittsburgh Courier*, June 13 2024, https://newpittsburghcourier.com/2024/06/13/a-courier-special-report-some-30-years-ago-wamo-began-juneteenth-celebrations-in-pittsburgh/

Pittsburgh Steps

Bob Regan, *Pittsburgh Steps, The Story of the City's Public Stairways*. Globe Pequot, 2015. ISBN 978-1-4930-1384-5.

City of Pittsburgh, Pittsburgh Citywide Steps Assessment, 2018, https://pittsburghpa.gov/domi/city-steps/

City of Pittsburgh, Steps Dashboard, https://pittsburghpa.maps.arcgis.com/apps/dashboards/84a0e248c87f4712baa1afd-836f3ea4f

Oliver Morrison, Pittsburgh's city steps to get $7 million upgrade, WESA, January 23 2023, https://www.wesa.fm/development-transportation/2023-01-23/pittsburghs-city-steps-to-get-7-million-upgrade

Ride the Trolley

Pittsburgh Railways Transit Guide, 1954, https://transitguru.info/pgh/history/pdfs/1954a.pdf

Pittsburgh PA and Vicinity Street and Interurban Railway Trackage 1859-1959, 1959, https://transitguru.info/pgh/history/pdfs/1959.pdf

Brookline Connection, History of Pittsburgh Inclines, https://www.brooklineconnection.com/history/Facts/Inclines.html

Ed Blazina, Finally: Bus Rapid Transit construction between Oakland and Downtown to begin Sept. 13, Pittsburgh Union Progress, September 1 2023, https://www.unionprogress.com/2023/09/01/finally-bus-rapid-transit-construction-between-oakland-and-downtown-pittsburgh-to-begin-sept-13/

Pittsburgh Regional Transit, University Line Service, https://www.rideprt.org/inside-Pittsburgh-Regional-Transit/projects-and-programs/bus-rapid-transit/BRT-service/

Bike Lanes

City of Pittsburgh, Bike(+) Master Plan, 2020, https://apps.pittsburghpa.gov/redtail/images/9994_Pittsburgh_Bike+_Plan_06_15_2020.pdf

BikePGH, Bicycling and Walking in Pittsburgh 2022, https://bikepgh.org/wp-content/uploads/2022/02/Cycling-Trends-Snapshot-in-Pittsburgh-2022.pdf

POGOH, https://pogoh.com/

Pittsburgh Marathon

Pittsburgh Marathon, https://www.thepittsburghmarathon.com/

2024 Pittsburgh Marathon facts and figures, *TribLive*, May 3 2024, https://triblive.com/sports/2024-pittsburgh-marathon-facts-and-figures/

Karen Price, A microwave oven on a sidewalk detours Pittsburgh marathon, *TribLive*, May 3 2010, https://archive.triblive.com/news/a-microwave-oven-on-a-sidewalk-detours-pittsburgh-marathon/

The Parking Spot Outside the Evergreen Cafe on Penn Avenue

Hannah Kinney-Kobre, EXCLUSIVE: Evergreen Cafe parking menace breaks silence after 50 years of pissing off Penn Avenue motorists, *Pittsburgh City Paper*, April 17 2023, https://www.pghcitypaper.com/news/exclusive-evergreen-cafe-parking-menace-breaks-silence-after-50-years-of-pissing-off-penn-avenue-motorists-23692874

Ali Trachta, Evergreen Cafe owner's car is still out front, now just "loading" in the loading zone every half hour, *Pittsburgh City Paper*, March 1 2024, https://www.pghcitypaper.com/news/evergreen-cafe-owners-car-is-still-out-front-now-just-loading-in-the-loading-zone-every-half-hour-25525640

Matt Petras, Both the Evergreen Cafe owner and his adversaries are calling authorities over the new loading zone, *Pittsburgh City Paper*, March 28 2024, https://www.pghcitypaper.com/news/evergreen-cafe-owner-and-penn-ave-parking-menace-phil-bacharach-and-his-adversaries-are-calling-authorities-over-the-new-loading-zone-25676268

Matt Petras, A petition for the Evergreen Cafe owner to get parking back is slowly racking up signatures on Change.org, *Pittsburgh City Paper*, May 6 2024, https://www.pghcitypaper.com/news/a-petition-for-the-evergreen-cafe-owner-to-get-parking-back-is-slowly-racking-up-signatures-on-changeorg-25905705

Section Three: Communities and Neighborhoods

Mount Oliver

Katie Blackley, How Mt. Oliver Borough eluded City of Pittsburgh annexation, WESA, March 13 2017, https://www.wesa.fm/arts-sports-culture/2017-03-13/how-mt-oliver-borough-eluded-city-of-pittsburgh-annexation

About Mt. Oliver, https://mtoliver.com/about/

US Census, https://data.census.gov/

Gentrification

US Census, https://data.census.gov/

J.A. Hirsch and L.H. Schinasi, A measure of gentrification for use in longitudinal public health studies based in the United States, Drexel University Urban Health Collaborative, 2019, https://drexel.edu/~/media/Files/uhc/briefs/Gentrification_Brief.ashx?la=en

Ryan Deto, Pittsburgh is one of the most gentrified cities in the U.S., *Pittsburgh City Paper*, April 4 2019, https://www.pghcitypaper.com/news/pittsburgh-is-one-of-the-most-gentrified-cities-in-the-us-14381722

Julia Felton, Pittsburgh Land Bank completed 1st sales in 2023, looks to ramp up work in new year, *TribLive*, December 8 2023, https://triblive.com/local/pittsburgh-land-bank-completed-first-sales-in-2023-looks-to-ramp-up-work-in-new-year/

Eric Jankiewicz, Penguins-picked developers make now-or-maybe-never pitch for Hill District land, *Public Source*, April 22 2023, https://www.publicsource.org/penguins-development-lower-hill-district-pittsburgh-block-e-economy/

Bill O'Driscoll, Removed Billboard Art Sparks Controversy, Public Forum, WESA, April 5 2018, https://www.wesa.fm/arts-sports-culture/2018-04-05/removed-billboard-art-sparks-controversy-public-forum

Older Houses and Lead Paint

US Census, https://data.census.gov/

Wikipedia, National Register of Historic Places listings in Pittsburgh, Pennsylvania, https://en.wikipedia.org/wiki/National_Register_of_Historic_Places_listings_in_Pittsburgh,_Pennsylvania

Habitat for Humanity of Greater Pittsburgh, Why Housing Matters, https://www.habitatpittsburgh.org/why-housing-matters

Bob Bauder, Residents living in history as Pittsburgh preps for bicentennial, *TribLive*, June 23 2016, https://archive.triblive.com/local/pittsburgh-allegheny/residents-living-in-history-as-pittsburgh-preps-for-bicentennial/

Allegheny County Health Department, Lead Data, https://www.alleghenycounty.us/Services/Health-Department/Lead-Exposure-Prevention/Lead-Data

Children and Seniors

US Census, https://data.census.gov/

Lajja Mistry, While Pittsburgh's district bleeds students, a few schools grow, *Public Source*, January 9 2024, https://www.publicsource.org/pittsburgh-public-schools-pps-uneven-scales-uprep-perry-enrollment-success/

Scott R. Beach, Christopher Briem, Janet Schlarb, Everette James, and Meredith Highes, State of Aging, Disability, & Family Caregiving in Allegheny County, University of Pittsburgh Center for Social & Urban Research, 2022, https://ucsur.pitt.edu/state_of_aging_2022.php

Jillian Forstadt, Pittsburgh Public Schools to hold 10 town halls on building utilization proposal this summer, WESA, June 12 2024, https://www.wesa.fm/education/2024-06-12/pittsburgh-public-schools-to-hold-10-town-halls-on-building-utilization-proposal-this-summer

Race and Ethnicity

US Census, https://data.census.gov/

Othering & Belonging Institute, Most to least segregated cities in 2020, University of California at Berkeley, https://belonging.berkeley.edu/most-least-segregated-cities-in-2020

Western Pennsylvania Regional Data Center, Redlining Maps from the Home Owners Loan Corporation, 1937, https://data.wprdc.org/dataset/redlining-maps-from-the-home-owners-loan-corporation

WESA, 50 Years After The Fair Housing Act, Buyers And Renters Still Face Discrimination In Pittsburgh, WESA, April 11 2018, https://www.wesa.fm/identity-justice/2018-04-11/50-years-after-the-fair-housing-act-buyers-and-renters-still-face-discrimination-in-pittsburgh

Tree Cover

PA Spatial Data Access, Allegheny County – Urban Tree Canopy, https://www.pasda.psu.edu/uci/DataSummary.aspx?dataset=1203

Justin Stewart, What's next for Hays Woods, Pittsburgh's newest park, *Allegheny Front*, September 8 2023, https://www.alleghenyfront.org/hays-woods-pittsburgh-park-invasive-plants/

Tree Pittsburgh, https://www.treepittsburgh.org

Matt Petras, Pittsburgh's uneven canopy begs question: Can the city achieve tree equity?, *Public Source*, May 25, 2022, https://www.publicsource.org/pittsburgh-tree-canopy-equity-hazelwood-hill-district-hilltop-climate-action-shade/

Supermarkets and Food Deserts

Southwestern Pennsylvania Commission, Supermarkets and Other Grocery (except Convenience) Stores, 2020, https://hub.arcgis.com/datasets/5ee3f02621c545ac8caef5b5e5be533c/explore?location=40.444916 percent2C-79.909583 percent2C11.55

Shelly Danko+Day, FeedPGH: Understanding Food Insecurity in Pittsburgh, 2019, https://apps.pittsburghpa.gov/redtail/images/16669_FeedPGH_Print_Version_11.18.21.pdf

Lauren Linder, Salem's Market opens in Pittsburgh's Hill District, ending food desert, KDKA, February 8 2024, https://www.cbsnews.com/pittsburgh/news/salems-market-pittsburgh-hill-district-grand-opening/

Talia Kirkland, Hazelwood church plans to operate grocery co-op, WPXI, October 12 2023, https://www.wpxi.com/news/local/hazelwood-church-plans-operate-grocery-co-op/EYU7JXHF2ZCZRAUBWVHUBFE65U/

Chinatown

Chris Potter, Was there an Asian influence in Pittsburgh's history?, *Pittsburgh City Paper*, March 9 2006, https://www.pghcitypaper.com/columns/was-there-an-asian-influence-in-pittsburghs-history-1334237

Woodene Merriman, Inn to the past: Downtown Cantonese restaurant points back to city's vanished Chinatown, *Post-Gazette*, December 9 2003, https://web.archive.org/web/20120516223514/http://old.post-gazette.com/lifestyle/20031209chinatown1209p1.asp

Jordana Rosenfeld, Downtown Pittsburgh celebrates Chinatown's official recognition as historic landmark, *Pittsburgh City Paper*, April 13 2022, https://www.pghcitypaper.com/news/downtown-pittsburgh-celebrates-chinatowns-official-recognition-as-historic-landmark-21467830

AR Mixtory, Pittsburgh Chinatown, https://projects.etc.cmu.edu/ar-mixtory/pittsburgh-chinatown/

Kathleen J. Davis, What happened to Pittsburgh's Chinatown? WESA, February 14 2019, https://www.wesa.fm/arts-sports-culture/2019-02-14/what-happened-to-pittsburghs-chinatown

Lena Chen, Uncovering Pittsburgh's long-hidden Asian American history made me feel at home — and I learned I wasn't alone, *Public Source*, July 28 2023, https://www.publicsource.org/pittsburgh-mayor-chinatown-asian-americans-film-last/

Immigrant Population

US Census, https://data.census.gov/

Jillian Forstadt, More refugees are resettling in Allegheny County than ever before, agencies say, WESA, December 26, 2022, https://www.wesa.fm/identity-community/2022-12-26/allegheny-county-saw-historic-influxes-in-refugees-this-year-and-more

City of Pittsburgh, Welcoming Pittsburgh, https://pittsburghpa.gov/wp/

LGBTQ+Burgh

QBurgh, https://qburgh.com/
Ollie Gratzinger, Pride Month: The History of Pittsburgh's Gay Pride Parade Spans Decades, *Pittsburgh Magazine*, June 7 2022, https://www.pittsburghmagazine.com/pride-month-the-history-of-pittsburghs-gay-pride-parade-spans-decades/

Jewish Pittsburgh

Jewish Federation of Greater Pittsburgh, https://jewishpgh.org/info/synagogues/
Pittsburgh Eruv, https://www.pittsburgheruv.org/
Chris Hedlin, How did Jewish life in Pittsburgh end up centered in Squirrel Hill?, *Public Source*, January 31 2022, https://www.publicsource.org/pittsburgh-faith-race-place-jewish-life-centered-in-squirrel-hill/

Catholic Pittsburgh

Diocese of Pittsburgh, List of Current Churches, https://diopitt.org/list-of-current-churches
Wikipedia, List of churches in the Roman Catholic Diocese of Pittsburgh, https://en.wikipedia.org/wiki/List_of_churches_in_the_Roman_Catholic_Diocese_of_Pittsburgh
Rachel Wilkinson, A Pittsburgh Church Holds the Greatest Collection of Relics Outside of the Vatican, *Smithsonian Magazine*, July 2017, https://www.smithsonianmag.com/arts-culture/pittsburgh-church-greatest-collection-relics-outside-vatican-180963680/

Muslim Pittsburgh

Chris Hedlin, Pittsburgh was once a Black Muslim refuge. Here's one family's story., *Public Source*, May 6 2021, https://www.publicsource.org/abdullah-family-black-muslim-pittsburgh-history/
Chris Hedlin, Who were Pittsburgh's earliest Muslims? Who built the city's first mosque?, *Public Source*, January 31 2022, https://www.publicsource.org/pittsburgh-faith-race-place-earliest-muslim-history/
David S. Rotenstein, Saeed Akmal stepped out of his brother's shadow to build Pittsburgh's Black Muslim community, *Next Pittsburgh*, December 5 2022, https://nextpittsburgh.com/pittsburgh-for-all/meet-saaed-akmal-the-community-builder-from-the-bellinger-family/
Haroon Al-Qahtani, An Oral History of Islam in Pittsburgh, 2006, https://archive.org/details/An_Oral_History_of_Islam_in_Pittsburgh/an_oral_history_of_islam_in_pittsburgh-hi.mp4

Mexican War Streets

Katie Blackley, How the Mexican War Streets got its name, WESA, May 5 2022, https://www.wesa.fm/arts-sports-culture/2022-05-05/how-the-mexican-war-streets-got-its-name
Chris Potter, What's the historical significance of the "Mexican War Streets" on the North Side?, *Pittsburgh City Paper*, June 12 2003, https://www.pghcitypaper.com/columns/whats-the-historical-significance-of-the-mexican-war-streets-on-the-north-side-1335170

Blue Dot in a Red Sea

Allegheny County Election Results, https://www.alleghenycounty.us/Government/Elections/Election-Results
Oliver Morrison, How Allegheny County delivered Pennsylvania to Biden, *Public Source*, November 12, 2020, https://www.publicsource.org/biden-trump-allegheny-county-pittsburgh-vote-breakdown/
Chris Potter, Summer Lee defeats Bhavini Patel in Democratic primary for 12th Congressional District, WESA, April 23, 2024, https://www.wesa.fm/politics-government/2024-04-23/pa-election-results-congress-12th-district

To Outer Space

Sarah Abrams, Astronaut Jay Apt's career launched—and landed—in Pittsburgh, *Pittsburgh Jewish Chronicle*, July 14, 2021, https://jewishchronicle.timesofisrael.com/astronaut-jay-apts-career-launched-and-landed-in-pittsburgh/
Patrick Damp, NASA astronaut and Pittsburgh native Warren 'Woody' Hoburg takes Terrible Towel to space, KDKA, September 2 2023, https://www.cbsnews.com/pittsburgh/news/nasa-astronaut-and-pittsburgh-native-warren-woody-hoburg-takes-terrible-towel-to-space/
Joe Napsha, Kecksburg celebrates UFO Festival, attracting believers, others, *TribLive*, July 20 2024, https://triblive.com/local/westmoreland/kecksburg-celebrates-ufo-festival-this-weekend-attracting-believers-and-those-who-just-like-food-games-and-fun/
Alex Fitzpatrick, Erin Davis, and Alice Feng, Pennsylvania doesn't report many UFO sightings, Axios, February 13, 2024, https://www.axios.com/local/philadelphia/2024/02/13/pennsylvania-ufo-sightings-map
Ralph W. Stone and Eileen M. Starr, Meteorites found in Pennsylvania, Pennsylvania Geological Survey, 1967, https://repository.tcu.edu/bitstream/handle/116099117/57720/Black_Moshannon_001.pdf

Section Four: Places in the City

The Cathedral of Learning
Wikipedia, Nationality Rooms, https://en.wikipedia.org/wiki/Nationality_Rooms

University of Pittsburgh Campus Tour, Cathedral of Learning, https://www.tour.pitt.edu/tour/cathedral-learning

Mister Rogers' Neighborhood
Wikipedia, Fred Rogers, https://en.wikipedia.org/wiki/Fred_Rogers

Visit PA, Taking a Stroll in Fred Rogers' Neighborhood, https://www.visitpa.com/article/taking-stroll-fred-rogers-neighborhood

Rob Owen and Barbara Vancheri, Fred Rogers dies at 74, *Pittsburgh Post-Gazette*, February 28 2003, https://web.archive.org/web/20230410134057/https://old.post-gazette.com/ae/20030228rogersae1p1.asp

City of Dinosaurs
Margaret Fleming, Which Dinosaurs lived in Pittsburgh?, WESA, August 31 2021, https://www.wesa.fm/arts-sports-culture/2021-08-31/dinosaurs-pittsburgh

Chandni Patel and Danielle Grentz, Carnegie and his Dinosaur Fascination, *Secret Pittsburgh*, https://secretpittsburgh.pitt.edu/sp/node/315

Dinomite Days, https://dinomitedays.org/index.htm

Carnegie Museum of Natural History, Jurassic Days: Dino Statue Driving Tour, https://carnegiemnh.org/jurassic-days-dino-statue-driving-tour/

Shelby Brokaw, DinoMite Days: When Dinosaurs ruled Pittsburgh, The Theme Park Files, April 15 2021, https://www.thethemeparkfiles.com/blog/dino-mite-days-when-dinosaurs-ruled-pittsburgh

Carnegie's Legacy
Wikipedia, Andrew Carnegie, https://en.wikipedia.org/wiki/Andrew_Carnegie

Susan Stamberg, How Andrew Carnegie Turned His Fortune Into A Library Legacy, NPR, August 1 2013, https://www.npr.org/2013/08/01/207272849/how-andrew-carnegie-turned-his-fortune-into-a-library-legacy

Colin Alexander, Expert blog: Andrew Carnegie and the Great Philanthropic Misunderstanding, Nottingham Trent University, March 26 2021, https://www.ntu.ac.uk/about-us/news/news-articles/2021/03/expert-blog-andrew-carnegie-and-the-great-philanthropic-misunderstanding

Steel City
Rivers of Steel National Heritage Area, https://riversofsteel.com

Jamie Wiggan, From steelworkers to baristas: the new face of Pittsburgh's evolving labor movement, *Pittsburgh City Paper*, August 31 2022, https://www.pghcitypaper.com/news/from-steelworkers-to-baristas-the-new-face-of-pittsburghs-evolving-labor-movement-22328278

Jack Troy, U.S. Steel closes portion of Clairton Coke Works as part of settlement, *TribLive*, June 6 2024, https://triblive.com/local/regional/u-s-steel-closes-portion-of-clairton-coke-works-as-part-of-settlement/

Air Quality
US EPA, Envirofacts, https://enviro.epa.gov/

Allegheny County Health Department, Air Quality Annual Report 2022, https://www.alleghenycounty.us/files/assets/county/v/1/government/health/documents/air-quality/reports/aq-annual-report/2022-air-quality-annual-report.pdf

American Lung Association, State of the Air 2024, https://www.lung.org/getmedia/dabac59e-963b-4e9b-bf0f-73615b07bfd8/State-of-the-Air-2024.pdf

Doug Shugarts and Sarah Boden, Air quality declines in Pittsburgh as smoke and haze from Canadian wildfires blanket metro area, WESA, June 28 2023, https://www.wesa.fm/environment-energy/2023-06-28/pittsburgh-air-canadian-wildfires

Centers for Disease Control and Prevention, PLACES dataset, 2023, https://chronicdata.cdc.gov/500-Cities-Places/PLACES-Census-Tract-Data-GIS-Friendly-Format-2023/yjkw-uj5s/about_data

Allegheny County Health Department, Asthma Task Force Report, 2019, https://www.alleghenycounty.us/files/assets/county/v/1/government/health/documents/2019-asthma-task-force-report.pdf

Reed Frazier, Study: Pittsburgh kids near polluting sites have higher asthma rates, State Impact Pennsylvania, November 11 2020, https://stateimpact.npr.org/pennsylvania/2020/11/11/study-pittsburgh-kids-near-polluting-sites-have-higher-asthma-rates/

Fracking and Sunning
PA Department of Environmental Protection, Oil and Gas Well Inventory, https://greenport.pa.gov/ReportExtracts/OG/OilGasWellInventoryReport

Pennsylvania Solar Center, https://pasolarcenter.org/resources/solar-installations/

Rachel Carson
Wikipedia, Rachel Carson, https://en.wikipedia.org/wiki/Rachel_Carson

Michael B. Smith, Silence, Miss Carson! science, gender, and the reception of "Silent Spring", *Feminist Studies*, 2001.

WITF, Rachel Carson: Voice of Nature, 2018, https://www.pbs.org/video/rachel-carson-voice-of-nature-viqzrt/

August Wilson
Wikipedia, August Wilson, https://en.wikipedia.org/wiki/August_Wilson

WQED, August Wilson: The Ground on Which I Stand, American Masters, February 20 2015, https://www.pbs.org/wnet/americanmasters/august-wilson-the-ground-on-which-i-stand-about-the-film/3610/

WQED, August Wilson's Hill District, https://www.wqed.org/august-wilson/hill-district-map/

John L. Dorman, August Wilson's Pittsburgh, *New York Times*, August 15 2017, https://www.nytimes.com/2017/08/15/travel/august-wilsons-pittsburgh.html

The Sporting Life
Wikipedia, Sports in Pittsburgh, https://en.wikipedia.org/wiki/Sports_in_Pittsburgh

Eric Jankiewicz, Penguins-picked developers make now-or-maybe-never pitch for Hill District land, *Public Source*, April 22 2023, https://www.publicsource.org/penguins-development-lower-hill-district-pittsburgh-block-e-economy/

Julia Fraser, Construction on new concert venue in Pittsburgh's Lower Hill could begin this fall, WESA, June 13 2024, https://www.wesa.fm/development-transportation/2024-06-13/pittsburgh-hill-district-concert-venue-live-nation-penguins

The Immaculate Reception
Wikipedia, Immaculate Reception, https://en.wikipedia.org/wiki/Immaculate_Reception

Heinz History Center, A Pivotal Moment – December 23, 1972: The Immaculate Reception, https://www.heinzhistorycenter.org/blog/western-pennsylvania-history-a-pivotal-moment-immaculate-reception/

Gary Mihoces, What is 'The Immaculate Reception'? The controversy and how the famous NFL play got its name, *USA Today*, December 21 2022, https://www.usatoday.com/story/sports/nfl/steelers/2022/12/21/immaculate-reception-how-famous-nfl-play-got-its-name/10937196002/

Acknowledgments

When I interview authors for the *New Books Network* podcast, they often tell me they hardly know where to begin in thanking everyone who helped them along the way, and now that I'm finishing my own book I know what they mean. I'll begin by thanking Monks and Rodrigo for keeping me safe from bugs and loud cars while working on this book. This book would not have been remotely possible without the many people in the city, county, state, and federal governments compiling data and making it easily available. As a former local news employee, I give a special salute to local reporters—at WESA, *Public Source*, *City Paper*, *Pittsburgh Union Progress*, *TribLive*, and the *New Pittsburgh Courier*—for their invaluable work in giving context for the many issues I addressed in this book (wherever possible I avoided citing the *Post-Gazette*, in solidarity with the reporters in the Newspaper Guild who have been on strike since October 2022). A general thank-you goes to everyone I have met in Pittsburgh—yinz have showed me so many fascinating nooks and crannies of our city. Appreciation also goes to the team at Belt Publishing, especially Anne Trubek, Hattie Fletcher and Michael Jauchen. Finally, my thanks to Christina, Sara, and Suzanne for keeping me sane through this process.

About the Author

Stentor Danielson has been a map nerd since reading *The Lord of the Rings* in 3rd grade. After spending their childhood in both western and eastern Pennsylvania, they earned a BA in Geography from Colgate University and a PhD in Geography from Clark University. Since 2009 they have lived in (or just outside) Pittsburgh while teaching geography and environmental studies at Slippery Rock University. Since 2015 they have run Mapsburgh, an online shop making paper cutout street maps and fantasy-style maps of real places. They currently share a house with one human partner, two feline bosses, and a classic mid-century Pittsburgh pink tile bathroom.